THE FLUORSPAR MINES OF NEWFOUNDLAND

McGILL-QUEEN'S/ASSOCIATED MEDICAL SERVICES STUDIES IN THE HISTORY OF MEDICINE, HEALTH, AND SOCIETY

SERIES EDITORS: S.O. FREEDMAN AND J.T.H. CONNOR

Volumes in this series have financial support from Associated Medical Services, Inc. (AMS). Associated Medical Services Inc. was established in 1936 by Dr Jason Hannah as a pioneer prepaid not-for-profit health care organization in Ontario. With the advent of medicare, AMS became a charitable organization supporting innovations in academic medicine and health services, specifically the history of medicine and health care, as well as innovations in health professional education and bioethics.

1 Home Medicine
The Newfoundland Experience
John K. Crellin

2 A Long Way from Home
The Tuberculosis Epidemic among the Inuit
Pat Sandiford Grygier

3 Labrador Odyssey
The Journal and Photographs of Eliot Curwen on the Second Voyage of Wilfred Grenfell, 1893
Ronald Rompkey

4 Architecture in the Family Way
Doctors, Houses, and Women, 1870–1900
Annmarie Adams

5 Local Hospitals in Ancien Régime France
Rationalization, Resistance, Renewal, 1530–1789
Daniel Hickey

6 Foisted upon the Government?
State Responsibilities, Family Obligations, and the Care of the Dependent Aged in Nineteenth-Century Ontario
Edgar-André Montigny

7 A Young Man's Benefit
The Independent Order of Odd Fellows and Sickness Insurance in the United States and Canada, 1860–1929
George Emery and J.C. Herbert Emery

8 The Weariness, the Fever, and the Fret
The Campaign against Tuberculosis in Canada, 1900–1950
Katherine McCuaig

9 The War Diary of Clare Gass, 1915–1918
Edited by Susan Mann

10 Committed to the State Asylum
Insanity and Society in Nineteenth-Century Quebec and Ontario
James E. Moran

11 Jessie Luther at the Grenfell Mission
Edited by Ronald Rompkey

12 Negotiating Disease
Power and Cancer Care, 1900–1950
Barbara Clow

13 For Patients of Moderate Means
A Social History of the Voluntary Public General Hospital in Canada, 1890–1950
David Gagan and Rosemary Gagan

14 Into the House of Old
A History of Residential Care in British Columbia
Megan J. Davies

15 St Mary's
The History of a London Teaching Hospital
E.A. Heaman

16 Women, Health, and Nation
Canada and the United States since 1945
Edited by Georgina Feldberg, Molly Ladd-Taylor, Alison Li, and Kathryn McPherson

17 The Labrador Memoir of Dr Henry Paddon, 1912–1938
Edited by Ronald Rompkey

18 J.B. Collip and the Development of Medical Research in Canada
Extracts and Enterprise
Alison Li

19 The Ontario Cancer Institute
Successes and Reverses at Sherbourne Street
E.A. McCulloch

20 Island Doctor
John Mackieson and Medicine in Nineteenth-Century Prince Edward Island
David A.E. Shephard

21 The Struggle to Serve
A History of the Moncton Hospital, 1895 to 1953
W.G. Godfrey

22 An Element of Hope
Radium and the Response to Cancer in Canada, 1900–1940
Charles Hayter

23 Labour in the Laboratory
Medical Laboratory Workers in the Maritimes, 1900–1950
Peter L. Twohig

24 Rockefeller Foundation Funding and Medical Education in Toronto, Montreal, and Halifax
Marianne P. Fedunkiw

25 Push!
The Struggle for Midwifery in Ontario
Ivy Lynn Bourgeault

26 Mental Health and Canadian Society
Historical Perspectives
Edited by James Moran and David Wright

27 SARS in Context
Memory, History, and Policy
Edited by Jacalyn Duffin and Arthur Sweetman

28 Lyndhurst
Canada's First Rehabilitation Centre for People with Spinal Cord Injuries, 1945–1998
Geoffrey Reaume

29 J. Wendell Macleod
Saskatchewan's "Red Dean"
Louis Horlick

30 Who Killed the Queen?
The Story of a Community Hospital and How to Fix Public Health Care
Holly Dressel

31 Healing the World's Children
Interdisciplinary Perspectives on Health in the Twentieth Century
Edited by Cynthia Comacchio, Janet Golden, and George Weisz

32 A Canadian Surgeon in the Army of the Potomac
Francis M. Wafer
Edited by Cheryl A. Wells

33 A Sadly Troubled History
The Meanings of Suicide in the Modern Age
John Weaver

34 SARS Unmasked
Risk Communication of Pandemics and Influenza in Canada
Michael G. Tyshenko with assistance from Cathy Patterson

35 Tuberculosis Then and Now
Perspectives on the History of an Infectious Disease
Edited by Flurin Condrau and Michael Worboys

36 Caregiving on the Periphery
Historical Perspectives on Nursing and Midwifery in Canada
Edited by Myra Rutherdale

37 Infection of the Innocents
Wet Nurses, Infants, and Syphilis in France, 1780–1900
Joan Sherwood

38 The Fluorspar Mines of Newfoundland
Their History and the Epidemic of Radiation Lung Cancer
John R. Martin

The Fluorspar Mines of Newfoundland

Their History and the Epidemic of Radiation Lung Cancer

JOHN R. MARTIN

McGill-Queen's University Press
Montreal & Kingston • London • Ithaca

ISBN 978-0-7735-4039-2 (cloth)
ISBN 978-0-7735-4040-8 (paper)

Legal deposit third quarter 2012
Bibliothèque nationale du Québec

Printed in Canada on acid-free paper that is 100% ancient forest free (100% post-consumer recycled), processed chlorine free

This book has been published with the help of a grant from the Memorial University of Newfoundland's Publications Subvention Program.

McGill-Queen's University Press acknowledges the support of the Canada Council for the Arts for our publishing program. We also acknowledge the financial support of the Government of Canada through the Canada Book Fund for our publishing activities.

Library and Archives Canada Cataloguing in Publication

Martin, John R., 1922–
The fluorspar mines of Newfoundland: their history and the epidemic of radiation lung cancer/John R. Martin.

(McGill-Queen's/Associated Medical Services studies in the history of medicine, health, and society; 38)
Includes bibliographical references and index.
ISBN 978-0-7735-4039-2 (bound) – ISBN 978-0-7735-4040-8 (pbk.)

1. Fluorspar industry – Health aspects – Newfoundland and Labrador – St. Lawrence. 2. Fluorspar industry – Employees – Diseases – Newfoundland and Labrador – St. Lawrence. 3. Fluorspar industry – Employees – Health and hygiene – Newfoundland and Labrador – St. Lawrence. 4. Radon – Carcinogenicity – Newfoundland and Labrador – St. Lawrence. 5. Radon – Toxicology – Newfoundland and Labrador – St. Lawrence. 6. Miners – Health and hygiene – Newfoundland and Labrador – St. Lawrence. 7. Fluorspar industry – Newfoundland and Labrador – St. Lawrence – History. 8. Lungs – Cancer – Epidemiology. I. Title. II. Series: McGill-Queen's/Associated Medical Services studies in the history of medicine, health, and society; 38

HD7269.M61C36 2012 363.17'99 C2012-901852-X

This book was typeset by Interscript in 10.5/13 Sabon.

To the people of St Lawrence, past and present, in admiration for their courage and stoicism in the face of tragedy and disappointments.

Contents

Abbreviations

AA	Alcan Archives
ACGIH	American Conference of Governmental Industrial Hygienists
ALARA	As low as reasonably achievable
Alcan	Aluminium Company of Canada
ANF	American Newfoundland Fluorspar Company
ASARCO	American Smelting and Refining Company
BRINCO	British Newfoundland Corporation
CNTU	Confederation National Trade Unions
COLD	Chronic Obstructive Lung Disease
DNA	Deoxyribonucleic acid
DOSCO	Dominion Steel and Coal
HSIMS	Health Sciences Information and Media Service
ILO	International Labour Organization
IOCC	Iron Ore Company of Canada
MHA	Member, House of Assembly, Newfoundland and Labrador
NALCO	Newfoundland and Labrador Corporation
NDP	New Democratic Party
NHW	Department of National Health and Welfare, Canada
NIOSH	National Institute for Occupational Safety and Health
NIS	Newfoundland Information Services
NL	Newfoundland and Labrador
NMA	Newfoundland Medical Association, later Newfoundland and Labrador Medical Association
NS	Nova Scotia
OHS	Occupational Health and Safety

PC	Progressive Conservative or Privy Councillor
PHS	Public Health Service, United States
PLC	Private Limited Company, UK
RC	Royal Commission
SLWPU	St Lawrence Workers Protective Union
TDB	Trade Dispute Board
TLV	Threshold Limiting Value
WC	Workmen's Compensation
WCB	Workmen's Compensation Board
WHSCC	Workplace Health, Safety and Compensation Commission
WL	Working Level
WLM	Working Level Month

Acknowledgments

I am indebted to a host of associations and people for assistance in writing this book: to the Associated Medical Services Incorporated for financial support; to the staffs of the National Archives of Canada, the Provincial Archives of Newfoundland and Labrador, the Centre for Newfoundland Studies, and the Legislative Library of the Province of Newfoundland and Labrador for archival services. The last's extensive collection of press clippings saved me much labour. The staff, particularly Carolyn Morgan, also kindly photostatted the documents I wished to consult and assisted me in many other ways. Workplace Health, Safety and Compensation Commission (WHSCC) of the province of Newfoundland and Labrador, especially Mr Eric Bartlett, was extremely helpful in retrieving the statistics and information I requested. To the Aluminium Company of Canada (Alcan), now Rio Tinto Alcan (RTA), for granting me access to their invaluable files on Newfluor, their St Lawrence fluorspar mining subsidiary, especially to Dr Manoel Arruda for providing statistics.[1] Also to Dr David Roomes, health director RTA, and Nicole Hebert, manager of archives RTA, for providing a photograph of Dr Brent. To Ms Jean Loder of the St Lawrence Heritage Society for photographs of the St Lawrence mining operations, and to Mr A.A. Etchegary for identifying several. To Mr Brad Way, Newfoundland and Labrador

1 Alcan's archives were opened to me on the understanding that the company would have an opportunity to go through the manuscript before publishing. I agreed. Alcan had no comments to make on the work but requested that a rider be added absolving the company from responsibility for any opinions expressed in the book. This has been done.

Department of Natural Resources, for providing St Lawrence Fluorspar mining statistics. To Mr Eugene Ryan, Ms Sylvia Ficken, and Mr Terry Upshall, Health Sciences Information and Media Service (HSIMS), Faculty of Medicine, Memorial University of Newfoundland, for preparing the map of the Gulf of St Lawrence (Ms Ficken) and the photographs (Mr Upshall). To Dr John Evans for drawing my notice to the unpublished autobiography of C.K. Howse. To Dr W. Arsenault for drawing my attention to and providing a copy of Wendy Martin's *Once Upon a Mine*; to Dr Rick Rennie for drawing my attention to and providing a copy of the "Trist" report. To Dr Christopher Martin for bringing to my attention the experiences of the US uranium mines. To those who granted me interviews or provided information (see list at back of the book). To all of these associations and individuals, I express my deep appreciation for the trouble they went to in providing information or wise advice. Dr John Crellin, Dr Hazen Russell, Ms Adele Poynter, Ms Stephanie Harlick, and my two sons, Mr Richard and Dr Christopher Martin, reviewed the manuscript. I am indebted to them for making helpful comments, correcting my syntax, spotting errors, and performing many other services and advice. Neither they nor Rio Tinto Alcan are in any way responsible for the statements or opinions expressed in the book. Ms Stephanie Harlick and Ms Jennifer Seeman verified and edited the endnotes, a tedious task for which I am deeply grateful. I must also acknowledge with thanks the labour of Ms Seeman in preparing the index. I am indebted to my wife, Claire for proofreading and encouragement throughout the project and to Peter Y.P. So and Susan Yu of The UPS Store for so expertly typing the manuscript. The editorial talents of Dr Joanne Muzak and her meticulous scrutiny of the manuscript are also gratefully acknowledged. Finally, I am indebted to the Publications Subvention Program at the Memorial University of Newfoundland for providing the financial support necessary for the publication of this book. This support is gratefully acknowledged.

Preface

In his introduction to George Rosen's 1943 classic, *The History of Miners' Diseases: A Medical and Social Interpretation*, Henry Sigerist, a pioneer in medical social history, observed that occupational diseases differ from other diseases "not biologically but socially. They are the result of working conditions and therefore affect the working classes."[1] Sigerist had put his finger on the unique feature of occupational diseases – their class distribution. It is this fact that has so much influenced how occupational diseases are viewed and interpreted. At the time that Sigerist penned these words, the fluorspar miners at St Lawrence were unwittingly being exposed to an invisible lethal hazard, one that only became evident a decade later when alarming numbers of cases of lung cancer began to be diagnosed in the underground workforce. The authorities became suspicious that they might be facing what was then an unheard of event in non-uranium mines, an epidemic of lung cancer.

It took some years for the puzzled investigators to find the cause – radon from the granite bedrock seeping into the mines in the groundwater, and there, it and its progeny rising to hazardous levels because of inadequate ventilation.[2] It was alpha radiation from the radon progeny that was causing the lung cancer. The epidemic was to have profound consequences not only for the victims but for their families and communities. All of which makes a remarkable story.

When confronted with an unforeseen outbreak of cancer of the lung such as occurred in the St Lawrence fluorspar mines, the authorities are faced with two tasks. The first is to identify the hazard responsible for the cancer and then bring the levels of the hazard down to levels deemed to be safe and keep them there. The second task is to ensure that the victims of exposure to the hazard and their surviving dependents are adequately compensated. How these two duties were tackled in the case of the St Lawrence epidemic of lung cancer is a major theme of this book.

Naturally, the epidemic of lung cancer occupies a prominent position in this work. However, the other events in the mines' history are instructive, and, in a sense, the book can be read as a compendium of issues common to the mining industry. Furthermore, during the years covered by this book, 1933–2011, Newfoundland passed from being a de facto dependency of the UK to becoming a province of Canada, and from being an underdeveloped to a developed region. The political and economic events that occurred during this period were bound to have an impact on the operations of the St Lawrence fluorspar mines. Consequently, note is taken of these incidents. In summary, while primarily an occupational health study, this is a multifaceted book.

Anyone writing in the area of occupational health cannot fail to be struck by how much passion its issues rouse. This is hardly surprising since so much of occupational health is preventive, directed at drawing up and enforcing permissible safe exposure standards to workplace hazards. These can be matters of life and death, and contentious especially if there is any scientific uncertainty about their validity. Controversy and matters of life and death were certainly at stake in the St Lawrence fluorspar mines. At the time the alpha radiation hazard was discovered, standard setting for permissible exposure limits to radiation hazards was in its infancy. There were no regulations in Canada.[3] The Newfoundland government had to come up with its own set. It did so and based them upon those recommended in some quarters of the United States but made them more stringent.[4] However, it added the rider that actual exposure should be kept as low as "practicable." Naturally, with a multifactorial disease like lung cancer, it would be impossible to eliminate it entirely from the workforce, and cases continued to appear among miners hired after the installation of the extra ventilation systems.

When it comes to setting permissible exposure standards to risks, regulators have to take account of not only quantitative estimates of risk but also of economic and human considerations. Consequently, standards are a blend of these three elements. It is the weight to be given to each in setting permissible exposure standards that has generated so much controversy. As a result, two radically different perspectives on occupational health and safety have emerged. One stresses economic factors in drawing up standards; in other words, this perspective focuses on cost benefit analyses. For the other, human safety is paramount, and cost benefit analyses for determining safe exposure standards are "a moral outrage."[5] Mark Sagoff has proposed the terms "detached" and "engaged" to designate these two perspectives on occupational health issues.[6] The "engaged," by occupying the high moral ground, puts the "detached" on the defensive.

A person's occupation or profession influences which of the two perspectives, "engaged" or "detached," is adopted. The managerial and technical staffs of industry are largely recruited from university professional and applied sciences faculties and schools, as well as from technical colleges.[7] Not surprisingly, such white-collar workers tend to be in the "detached" camp. After all, it is not the managerial staff but the blue-collar workers who are exposed to industrial hazards, a pattern that has been labelled "the unequal distribution of hazards."[8] On the other hand, the social sciences and humanities departments of universities have much less of a stake in industry, and it is they and the media that supply so many recruits to the "engaged" camp. Some of the most trenchant criticisms of the adverse aspects of industrialization come from these quarters.[9]

My own point of view is that of an academic occupational health physician, who was also a consultant to the government after the epidemic of lung cancer had peaked. Inevitably, this position imposes a dispassionate tone on my discussion of events. My aim has been to tell what happened and to explain the reasons and the context in which events occurred. Context, among other considerations, refers to events in the social, political, and economic circumstances of the day and how they affected the St Lawrence fluorspar mines. It also entails comparisons with the other mines in the province. The St Lawrence fluorspar mines and uranium mines shared a common task – having to cope with hazardous levels of alpha radiation. The experiences of the uranium mines of the Colorado Plateau informed the regulations adopted in St Lawrence. The Colorado Plateau experiences help complete the background picture. At the St Lawrence mines, there is another important dimension to keep in mind – the emotional devastation and material loss experienced by those afflicted with or waiting to develop cancer of the lung, and families left without a husband or father. These tragic aspects of the story have been told by others who had personal contact with the victims or their families. The best known of which is Elliott Leyton's *Dying Hard: The Ravages of Industrial Carnage.*[10]

Readers should take note, I have followed the terminology in vogue during the existence of the St Lawrence fluorspar mines. For example, Newfoundland was the official name of what is now the Province of Newfoundland and Labrador. Workmen's Compensation Board (WCB) was the name for what is now Workplace Health, Safety and Compensation Commission (WHSCC). Readers are referred to the appendices for statistical data and to the glossary of technical terms for explanations of such terms.

While I was revising this book, Rick Rennie's *The Dirt: Industrial Disease and Conflict at St Lawrence, a History of the St Lawrence Fluorspar Mines* appeared.[11] There are enough differences between his and this work to justify another history of the mines.

John R. Martin

St Lawrence, 2000
(Photo by the author)

The Bay from Great St Lawrence Harbour, 2000
(Photo by the author)

Open Pit Iron Springs Mine, 1937
(Courtesy of St Lawrence Heritage Society)

Mill and Tailings St Lawrence Corporation
(Courtesy of Adele Poynter)

Director Mine, early years
(Courtesy of St Lawrence Heritage Society)

Director Mine, late years
(Courtesy of St Lawrence Heritage Society)

Above, below, and top of next page:
St Lawrence fluorspar miners. Note heavy rain gear.
(Courtesy of St Lawrence Heritage Society)

Early hoist
(Courtesy St Lawrence Heritage Society)

Cyril J. Walsh, MD
(Courtesy of Marian Clare Walsh)

Frank deN. Brent, MD
(Courtesy of Rio Tinto Alcan)

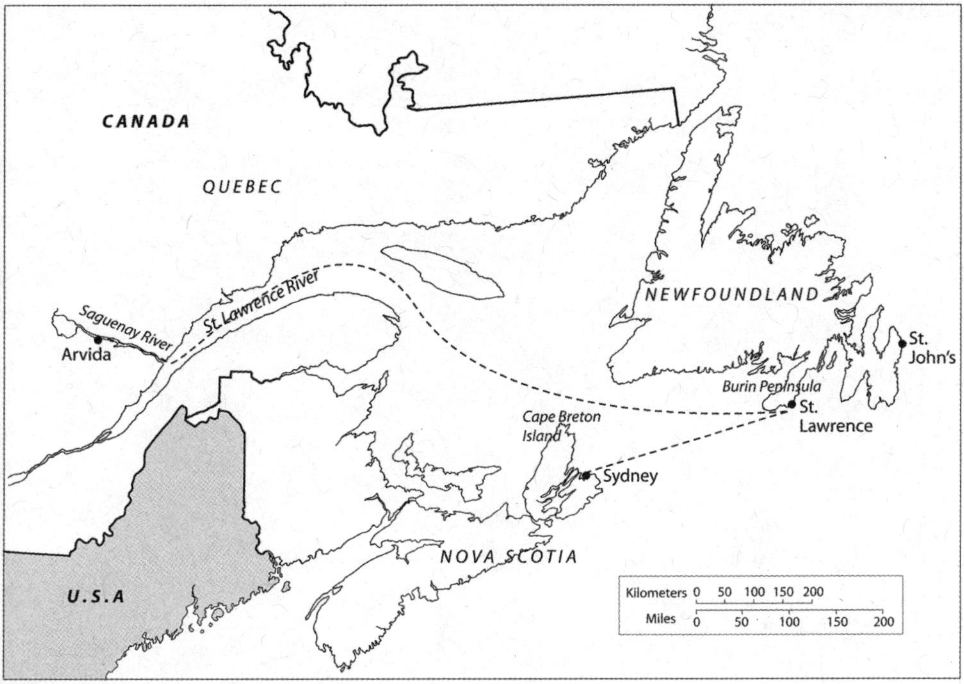

Newfoundland and Gulf of St Lawrence in the 1940s
(Prepared by Sylvia Ficken HSIMS)

THE FLUORSPAR MINES OF NEWFOUNDLAND

1

The St Lawrence Fluorspar Mines: The Early Days

In 1929, St Lawrence was a small Newfoundland "outport," one of a group of similar small fishing communities clustered around the tip of the Burin Peninsula, isolated from the rest of the world and reached only from the sea. The population of about 900 of mixed origin, mainly Irish but some English, supported itself by fishing. The year 1929 was to prove an eventful one in the life of the community. In that year, a tsunami, then called a tidal wave, swept into the town, destroying everything in its path. Not only was there loss of property and life, but fishing stocks – the only source of livelihood – disappeared, not to return for several years. In the world at large, the big event was the crash of the New York Stock Exchange, which triggered the worldwide Great Depression. This economic crash was to have a devastating impact on a community like St Lawrence, a community already so close to the poverty line. With their fish stocks gone, people were thrown onto the "dole" (relief), at six cents a day per person. This and what they could grow in their vegetable plots was all they had to get by on. Anyone who could show them a way out of the poverty prison would be greeted with open arms. Such a person was Walter E. Seibert, a self-assured flamboyant young accountant from New York.

Seibert assured the people of St Lawrence that they were sitting on top of a gold mine, figuratively speaking – the richest fluorspar deposit in the world; for such, he said, was the value of the lumps scattered on the ground around their town. All that was needed to get this torrent of riches flowing was to collect enough fluorspar to fill the hold of a schooner, ship it to Sydney, Cape Breton, Nova Scotia (NS), where it could be sold to Dominion Steel and Coal

(DOSCO) as a flux in steel smelting. Once this was done, and he was optimistic it could be done, it would open the door to the regular supply of fluorspar to DOSCO and other steel companies, bringing untold prosperity to an impoverished community.

The existence of fluorspar deposits in the St Lawrence area had been known since 1843.[1] They occur as inclined to vertical veins, embedded for the most part in granite and extending just over three kilometres. The fluorspar veins in St Lawrence show a "pinch and swell" pattern, that is, lenticular shaped deposits of the ore are joined at their narrow ends like beads in a necklace and measure up to thirty metres in width. About forty veins have been discovered but only about half have been suitable for mining. They have been categorized as high-grade – up to 95% calcium fluoride (CaF_2) with low amounts of contaminants and low-grade – up to 70% CaF_2 and as much as 50% silica (quartz).[2] Fluorspar had been used for centuries in the production of steel. When the steel mills opened on Cape Breton at the end of the nineteeth century, they used the local coal deposits, brought in iron ore from Bell Island, Newfoundland, but they had to import fluorspar from abroad. Thus, St Lawrence fluorspar had a potential market nearby. About this time, other industrial uses for fluorspar were developed, such as in the smelting of aluminum. The St Lawrence fluorspar deposits were now a valuable asset waiting to be put to commercial use. One of those who realized their potential worth was St John's butcher J. Campbell. In 1912, he obtained several leases at St Lawrence on land that contained fluorspar. He did nothing to develop his claims, and, in 1928, he sold them to W.H. Taylor also of St John's, and here Walter E. Seibert entered the picture.

Seibert's interest in the St Lawrence fluorspar deposits came about as the result of a series of chance happenings. In 1929, as a young accountant with the Corporation Trust Company of New York, he was sent to St John's in connection with some income tax business. While in St John's he met a relative of Taylor who put him in touch with the latter, presumably because Seibert had expressed an interest in the fluorspar deposits of St Lawrence. The two men met and eventually Taylor sold his claims to Seibert for $300. Seibert apparently bought the claims as a speculation because, subsequently, he bought up other claims. Finding it impossible to sell them, he turned to the idea of developing them himself. In 1931, he visited St Lawrence and stayed with Aubrey Farrell, a local merchant, who had been recommended to him as a reliable local representative. Together with Dr Warren S. Smith, an American geologist, he inspected his claims and made arrangements for six prospectors to explore them and select the most promising one for mining. Returning to New York, he was able to interest three partners in joining with him in putting up the capital for his venture. In November 1931, Seibert had his holdings incorporated as the St Lawrence Corporation (henceforth referred to as the Corporation) with a capital of $100,000 divided into 10,000 shares valued at ten dollars each and all issued to Seibert. He then turned to implementing his plan to collect enough fluorspar for a trial shipment to Sydney, NS, and there to be sold to DOSCO. Once again fortune smiled on him. He won the enthusiastic support of Father Thorne, the local parish priest – an essential ingredient for success in those days in a community like St Lawrence. Seibert then contracted with Farrell to collect 2,000 tons of fluorspar at four dollars per ton, enough to fill a schooner hold. Farrell was to supply the labour, to be paid at a rate of fifteen cents an hour. However, labourers would have to wait until the shipment was sold before collecting their wages. Local merchants agreed to advance credit until the men were paid. Seibert was to provide the equipment to excavate the fluorspar, which he did, bringing in what has been described as second-hand dilapidated equipment on the steamship *Renunga* in March 1933. On the strength of this agreement, Farrell hired twenty labourers. During the following months the men first built a mill house and then, using pick axes and jackhammers, toiled away at extracting the ore at an open pit that was later to become the Black Duck mine. The ore was transported by wheelbarrow in summer and by sled in winter to the mill house for crushing, washing, screening, and hand picking of the lumps of fluorspar. One man recounted a vivid boyhood memory of seeing sleds, loaded down with lumps of fluorspar, drawn by teams of oxen in tandem and wending their way to the mill.[3] By the spring of 1934, the agreed 2,000 tons of fluorspar had been amassed. It was then loaded into a schooner and taken to Sydney, NS. There it was tested to see if it met DOSCO's specifications for use as a flux in steel production. It passed the tests and was purchased for $24,000. The labourers were paid, and they in turn settled their accounts with the merchants. Seibert had proved his point. If DOSCO and other steady markets could be found for his fluorspar, St Lawrence could look forward to a bright future as a mining town.

Having demonstrated that St Lawrence fluorspar was marketable, Seibert moved on to placing his mining operations on a continuing basis with Dr Warren S. Smith in charge, counting on DOSCO as a steady and reliable customer.

THE OTHER NEWFOUNDLAND MINING COMPANIES

When the Corporation commenced operations there were two other mining companies in Newfoundland: the iron ore mines on Bell Island, Conception Bay, known as Wabana and owned by DOSCO; and a copper, lead, and zinc mine at Buchans, in the centre of the island, owned by the American Smelting and Refining Company (ASARCO).[4] Wabana shipped most of its ore to supply DOSCO's steel mill at Sydney. It ceased operations in 1966 chiefly because its product could no longer meet the specifications for steel manufacturing. ASARCO closed in 1984 because its mineral deposits were exhausted. Thus, the lifespans of the three mining companies overlap,[5] which permits comparison of practices and conditions at all of them.

The enthusiastic and cooperative spirit that produced and sold the first shipment of fluorspar was short-lived.[6] Looking back, those early days have been depicted in somewhat rosy terms – everyone working in unison to amass and sell fluorspar. The labourers, uncomplainingly putting up with low wages and delays in payment, and merchants advancing credit on paycheques, which could not be cashed until Seibert had made enough sales to honour them.[7]

VEINS EXPLOITED

Seibert invested his profits into further developing his operations and the purchase of equipment. The Black Duck, a high-grade mine, was developed first as an open pit, but in 1936 a shaft was sunk and underground mining commenced. It was shut down in 1941 because the fluorspar veins became too narrow to work profitably.[8] In 1935, work began at another high-grade mine, Iron Springs, first as an open pit and then, in 1938, underground. This was to be the major mine of the Corporation for the rest of its years. It exploited several other veins as open pits and one other underground mine – Blue Beach, also high-grade.[9] When it came to underground mining, drilling had hardly proceeded when a serious problem was encountered – one that was to plague underground mining at St Lawrence and to make it so costly. The problem was that the pits quickly flooded, and the greater the depth, the more rapidly this occurred. One witness described conditions at deeper levels as like a rainstorm.[10] Miners were constantly soaked.[11] Pushing wheelbarrows to the shaft through a foot of water and gravel was a Herculean task.[12] For mining to proceed, this constant deluge of water had to be controlled. It was estimated that 2,000 gallons of water a minute had to be pumped out of Iron Springs in its heyday, which accounted for three-quarters of the company's power consumption.[13] In using this much power, the Corporation encountered more troubles. It drew its power from a nearby hydroelectric plant. Even in the earliest days this system was insufficient to meet its needs. Consequently, it installed two 160-horsepower generators to provide supplementary power. In spite of this extra power, dry spells often reduced power to critical levels, forcing shutdowns.[14] Lack of capital dictated choice of equipment and mining methods.[15] Rennie Slaney, an erstwhile foreman with the Corporation, and Ena Farrell Edwards, co-author of a history of St Lawrence, have left vivid accounts of how appalling working conditions were in the mines in those days.[16] Miners used dry hammers operated by air compressors (jackhammers) slung over their shoulders, which were always chafed because of the constant friction. This method of drilling meant that the workers' faces were close to the rock face, and they breathed in the dust that their drilling generated. To protect themselves, the men would cover their noses and mouths with cheesecloth, which in no time became clogged. On coming up to the surface, miners would go through spasms of coughing until they had expectorated the dust they had inhaled. There were no safety lamps, helmets, toilets, dining rooms, or changing rooms with showers. The men had to climb down into the pits by ladder and walk to and from work. There was no ventilation other than what occurred naturally presumably because it was believed the wet conditions would abate dust. Yet there were constant complaints about dust escaping out of the drill holes. The mines operated on three consecutive eight-hour shifts, so there was never an opportunity for dust to settle. At first, the ore was loaded into wheelbarrows, which were pushed up an incline to the shaft. Later, primitive hand-operated hoists were used to pull carts. This description of working conditions in the 1930s sounds almost unbelievable, yet the men seem to have accepted their situation stoically and without complaint. Only a few chose to go back on the dole.[17]

There were always more applicants than available jobs.[18] Indeed, during the Corporation's lifetime the workforce was to remain stable with little turnover. Some who had started off with the company in 1933 were still with it in 1958.[19] As in primitive mining operations, it was not unusual for St Lawrence miners to keep a hand in their previous occupation – fishing – and even to withdraw from mining temporarily to give their full attention to fishing.[20]

The market for the Corporation's fluorspar was DOSCO and later other steel mills in eastern Canada. After a few years, the Corporation also began exporting to the United States.[21] The US market was to assume increasing importance. Production steadily rose and so did the size of the workforce (see Appendix B).

METHOD OF EMPLOYEE PAYMENT

The honeymoon between Seibert and the people of St Lawrence could not last forever. It was the method of delayed payment of wages that soured relations between Seibert and his employees.[22] The practice was to aim at issuing cheques every two weeks but there were often delays.[23] The cheques could not be cashed until the stock of fluorspar on hand had been sold, which could take months. In the meantime, an employee would take his cheque to a local store to make purchases. The merchant would retain it but issue IOUs for the unexpended portion of the cheque. These IOUs, written on torn-off portions of wrapping paper, circulated as currency. The merchants had to delay presenting their cheques at the Corporation's bank until they were assured there were sufficient funds to honour them. To their credit and patience, they bore the risks and delays in what was owing to them and did not charge interest. It was not only the labourers who had to wait to be paid. Dr Warren S. Smith, the manager, left after waiting for payment for two years.[24] Labourers were paid at fifteen cents an hour for an eight-hour shift, six days a week.[25] For some unexplained reason, $2.20 a month was held back as "back time" until 1936, when, by legal action and a threatened strike, the workers were able to put a stop to the practice.[26] By August 1939, wages had climbed to 23.3 cents per hour.[27] At the same time, ASARCO was paying thirty-seven to thirty-nine cents per hour, and Wabana was paying 32.36 cents per hour.[28] Furthermore, unlike Seibert, ASARCO and Wabana provided insurance schemes, and DOSCO provided low-cost housing for their Sydney workers.[29] Even though the three Newfoundland mining companies had different work schedules, overall, the financial and physical circumstances of workers at ASARCO and Wabana were superior to those at the Corporation because the mines at Buchans and Bell Island were owned by cartels (ASARCO and DOSCO, respectively), which had reserves of capital to spend on their holdings and subsidiaries. The Corporation, a lone company, had no such financial resources.

After several years, the authorities began to take notice of the shaky financial arrangements of the Corporation and the less than satisfactory working conditions. Its slowness in doing so must now be explained.

CHANGE OF GOVERNMENT

In 1926, the UK government informally conceded that the six self-governing dominions of the British Empire possessed full autonomy in external as well as in internal affairs, united only by allegiance to a common monarch. Legal recognition of the full independence of each dominion was proclaimed in 1931 by the Statute of Westminster. As the poorest and least populous of the dominions, coupled with its failure to pass the enabling legislation to take advantage of its constitutional rights, Newfoundland remained subordinate to the UK. Already heavily in debt, the stock market crash in 1929 was the coup de grâce to responsible government. Faced with impending bankruptcy and perhaps having to default payments on its bonds, the Newfoundland government turned to Canada and the UK for financial succour. Short-term loans were arranged, but the UK government warned the Newfoundland government that it would not be permitted to default on payments to its bondholders. However, the UK government would be willing to provide financial assistance in return for Newfoundland surrendering administrative and legislative authority to the UK government. The Newfoundland government had little choice but to agree, and the legislature voted itself out of existence. On 1 February 1934, the new administration, designated the Commission of Government, was installed. Newfoundland was reduced to the de facto status of a crown colony but, to spare feelings, was placed under the Office for Dominion Affairs. It was made clear by the UK government that this was a temporary arrangement. Once the island was able to stand on its feet financially, assurances were given that responsible government would be restored. Meanwhile, administrative and legislative authority was vested in a commission of six members appointed by the secretary of state for Dominion Affairs and chaired by the governor. Three of the commissioners were to be Newfoundlanders and three from the UK. Each headed up a department. What were considered to be the three most important departments – Finance, Public Utilities, and Natural Resources – were reserved for the UK representatives.

Under the new regime, oversight of mines was vested in the Division of Mines and the Geological Survey, a division first of the Department of Natural Resources, but some years later transferred to the Department of Public Utilities. The Geological Survey had been established in the first half of the nineteenth century, discontinued in 1912, revived in 1925 only to be shut down again in 1929.[30] However, ever since 1912, geological mapping and prospecting for minerals had been going on continuously in the summer months by post-graduate geology students from Princeton University under the supervision of their professors.[31]

In 1934, their supervisor was Professor A.K. Snelgrove. It was to him that the commissioner of Natural Resources, Sir John Hope-Simpson, turned for advice on reconstituting the Geological Survey and the regulation of the mines. Snelgrove was appointed government geologist in 1934. Since he was limited in the amount of time he could spend on the island, he stipulated that he be provided with a full-time associate government geologist. C.K. Howse was appointed to this post in 1934, and then, in 1935, he also became inspector of mines. In the 1930s, inspection of mines was performed by the government engineer, but it was only one of several of his tasks.[32] All he did was check for compliance with the Newfoundland Mining Code of 1916. Since neither he nor Howse, his immediate superior, were mining engineers, from time to time, a properly qualified mines inspector from the Ontario Department of Mines was brought in to conduct an expert inspection of the mines, focusing on safety, and to submit a report of his findings and recommendations to the commissioner, firstly of Natural Resources and later of Public Utilities.[33] In addition to his administrative duties, Howse began the geological mapping of the St Lawrence area as a PhD thesis at Princeton, but the outbreak of war put an end to this project.[34]

HOWSE'S INSPECTION VISITS

Howse paid his first visit of inspection to the Corporation in the summer of 1935. He was highly critical of what he saw. In his report, he summarized the site as "a small grossly underfinanced operation."[35] After receiving Howse's report, the commissioner of Natural Resources was concerned enough to put the matter of the Corporation before the Commission of Government. Citing Howse's report, the commissioner described the operation and what should be done to put the Corporation's affairs on a more satisfactory basis.[36] The Black Duck mine was a cleft in the rock, six feet wide and fifty feet deep, shored up by rough timbers, but Snelgrove and Howse deemed it safe. Of major concern was "the irregular payment of the miners." Seibert claimed he was making a profit of six dollars per ton.[37] The commissioner stated a viable profitable operation could be developed. What was needed was $150,000 capital to produce 25,000 tons fluorspar a year, which he estimated could be sold at nineteen to twenty-four dollars per ton. He urged the government to apply to the UK Colonial Development Fund for a loan of $150,000 to improve the operation and to sink a shaft to one hundred feet. The government would hold all the shares in the operation but would give Seibert a block of shares equal in value to what he had invested ($15,000) in the company, less his profits from sales. Seibert would also have the option of purchasing up to 49 per cent of the shares. If the government agreed, the commissioner would put this proposal to Seibert. Nothing seems to have come from his suggestion. Indeed, Bowering Company of New York had approached Seibert with a purchase offer for his St Lawrence holdings but had balked at his asking price of $500,000.[38]

On 1 June 1937, Howse revisited St Lawrence for the fourth time and reported his observations and recommendations to the commissioner of Natural Resources, R.B. Ewbank.[39] The workers had presented an ultimatum to Seibert on 20 April demanding a wage increase but so far had received no reply, or payment since 30 April. On 15 May, the men walked off the job. Management responded by closing the mine and began to hire boys as replacement workers. By the time of Howse's visit, the miners were in a desperate situation since they did not qualify for the dole. After five years of working for the Corporation, Howse observed miners were little if any better off than when they had been on the dole. The system of delayed payment was still in effect with Seibert issuing cheques in advance of funds to cover them. Worse still, workers were beginning to run up debts with the merchants who were beginning to object at advancing credit without security. Howse recommended firstly replacing the resident manager, C. Kelleher, whom he considered totally unsuited for the position. He also recommended an increase in the minimum wage to twenty-five cents an hour, payment in cash every two weeks. Finally, he encouraged Seibert to sell his St Lawrence holdings to an operator with sufficient capital to bring them up to acceptable standards. In his memorandum accompanying Howse's report, Ewbank had this to add: "I have seen the mine, and it is run in the most primitive way and is in the opinion of Mr Howse dangerous to those who work in it." He concluded by saying that he had no legal means for obliging Seibert to improve his holdings. However, he could call him in and "threaten him with an export tax unless he came forward with reasonable proposals." All that the government seems to have done was to write a letter to Seibert expressing dissatisfaction with his operations and asking what measures he proposed taking. Nothing seems to have come from this somewhat feeble démarche. Bad as things were at the Corporation, the government could not run the risk of its closure. Seibert knew this and indeed boasted that he could do just as he pleased.[40]

The Corporation's financial fortunes took a turn for the better in the late 1930s. Seibert obtained credit from a New York bank. The US tariff on imported fluorspar was lowered, which permitted fluorspar from St Lawrence to be traded at competitive prices on the US market.[41] With war in the offing and the anticipation of a big expansion of industries that required fluorspar, it seemed that the Corporation's financial troubles could be over, at least for the duration of the war. Meanwhile, others, probably stimulated by Seibert's success, began to wonder whether there might be room for more than one fluorspar mining company at St Lawrence.

AMERICAN NEWFOUNDLAND FLUORSPAR COMPANY

Seibert was not the only party eager to exploit the St Lawrence fluorspar deposits. In the 1920s and 1930s, over two dozen claims came into the hands of a consortium of St John's minor businessmen Hookey and Co., named after the senior partner, Joshua Hookey.[42] In 1931, a clash occurred with the Corporation when the latter staked out claims on land already claimed by Hookey and Co. The courts were called upon to adjudicate and ruled in favour of Hookey and Co. However, Hookey and Co. had neither the capital nor expertise to develop their claims and simply sat on them, something they were entitled to do indefinitely under the existing Mines Act of 1906.

Seibert's success in selling fluorspar to DOSCO in Sydney came to the notice of US steelmakers, generating an interest in importing St Lawrence fluorspar to supply their own needs. One of these businessmen was Edwin J. Lavino, president of E.J. Lavino Company of Philadelphia. He purchased 51 per cent ownership of Hookey and Co.'s claims and incorporated his holdings in May 1937 as the American Newfoundland Fluorspar (ANF) Company. The terms he dictated to Hookey and his associates were harsh. In return for Lavino putting up $10,000 capital, they had to relinquish participation in all mining policy and activities. Work began in the summer of 1937 on the Tarefare, a low-grade vein. Shortly afterwards, however, a more promising deposit, the Director vein, was discovered. Operations were transferred there in June 1938, and the Tarefare mine closed. At first, the Director mine was an open pit, but by 1939, a shaft 150 feet deep had been sunk.[43] Working conditions at the ANF Co. were far superior to those at the Corporation and the equipment more modern.[44] However, before the ANF Co. had shipped out any fluorspar, Lavino decided that the outbreak of war was a good time to sell.[45] He approached the Aluminium Company of Canada (Alcan) to see if it was interested in purchasing his company. After some bargaining, Alcan came up with an offer acceptable to Lavino. He called a stockholders' meeting and railroaded Alcan's offer, over the vehement protests of Hookey and his associates. The latter instituted legal action to prevent the transfer but this failed. In December 1939, Alcan formed a subsidiary, the Newfoundland Fluorspar Company (Newfluor) to operate its St Lawrence holdings.

NEWFLUOR

Newfluor concentrated its operations at the Director mine with its more accessible deposits and sunk another shaft.[46] Soon it ran into the same snag that beset Iron Springs mine – a constant deluge of water that required continuous pumping of 3,000 gallons a minute so that the men could work. Consequently, electricity bills were high ($150,000 a year). Even with supplementary diesel generators, power failures occurred, which necessitated shutdowns until power could be restored. The Director mine also ran into a problem that was not a feature of Iron Springs mine, namely rock bursts.[47] These also could be costly since the mine would have to be closed until it was deemed safe to resume work.

Newfluor continued to maintain the high standards of comfort for its workers set by the previous owner.[48] A changing house was provided containing a dining room, changing rooms, showers, and toilets. In contrast, the Corporation miners continued to eat in a drafty shack, relieve themselves where they could, and walk home in damp clothes.[49]

The reason Alcan bought the ANF Co. was that, with the onset of war, it was essential to have an assured and adequate source of fluorspar to supply its aluminum smelting plant at Arvida (now part of the city of Saguenay), at the top of the Saguenay River, Quebec. Newfluor satisfied both these requirements; it was what is known as a captive mine.[50] Such a subsidiary exists to supply adequate and assured supplies of a commodity required by the parent company for use in its industrial operations. The advantage of being a captive mine is that it can count on the parent company to invest in maintaining higher standards than it could afford on its own. However, a captive mine is vulnerable, lasting only as long as the parent company perceives benefits to itself in the arrangement. The Corporation, on the other hand, was an independent mine, owned by Seibert, selling all its product on the open market and subject to all its vagaries. In an independent mine, the amount of funds to be dispensed on maintenance and development of the operation as well as on working conditions, including health and safety, will depend on the size of the profits of the enterprise. In these respects, an independent mine is often at a disadvantage compared to a captive one. The difference between the Corporation and Newfluor, one an independent and the other a captive operation, explains much of their subsequent history.

Newfluor spent two years organizing and developing the Director mine before going into production. Soon production began to outstrip demand from Arvida. By 1945, the stockpile of fluorspar was so large that the mine had to be temporarily closed and flooded until the stockpile was exhausted.[51] For the Corporation the war years were to be ones of prosperity but also labour unrest.

2

The War Years and the Last Days of Commission of Government

It was the outbreak of war, more than anything else, that put the fortunes of the Corporation on a more solid basis. With access to European supplies of fluorspar closed, North America had to turn to sources within its own borders to supply its needs. In Canada, the war-time expansion of industries utilizing fluorspar created a great demand for the product. Should it get drawn into the conflict, the US government foresaw the need to be assured of reliable and sufficient sources of fluorspar.[1] Shortly after entering the war, the US government seems to have approached the Commission of Government explaining it would require 50,000 tons per annum of St Lawrence fluorspar. This would call for an expansion of the Corporation's plant and operations. To do so, the US War Production Board granted Seibert a defence plant loan to purchase heavy equipment and build a froth flotation plant to produce acid grade filter cake – the first of its kind in North America.[2] The filter cake would then be exported to Wilmington, Delaware, where Seibert had constructed a plant to process it further.[3] From this plant was to come the fluorspar used in the construction of the first atom bomb.[4] The Corporation continued to supply metallurgical grade fluorspar to eastern Canadian steel mills. To supply the increased amounts of fluorspar, another underground mine, Blue Beach, was opened as well as several open pits.

The outbreak of war had an impact on labour relations. The Commission of Government promulgated a number of laws and regulations that provided for the control of wages but not prices as well as for the reporting of unresolved labour disputes to the commissioner of Public Utilities before lockouts or strikes.[5] The commissioner then had several options – one being to appoint a board of inquiry or an arbitration tribunal with equal representation from the union and management, and a chairman selected by them or the commissioner. Another option, for the most intractable disputes, was the appointment of a trade dispute board (TDB). Regardless of which option he elected to follow, the commissioner had the ultimate authority to impose a settlement to a dispute; but he seems to have been reluctant to do so. The trade unions refused to forgo the option of striking for the duration of the war.[6] Consequently, during the war there were strikes and short-term work stoppages at all the mining companies except Newfluor.[7]

Working conditions improved slowly at the Corporation. Even by 1940, the Department of Public Health and Welfare was incensed at the deplorable sanitary conditions underground, Seibert's refusal to correct them, and the department's impotence to do anything about them.[8] Wet conditions continued to make work unpleasant.[9]

In the late 1930s and early 1940s, relations between management and workers became increasingly strained, culminating in five strikes in 1941. The last ended by the appointment of a trade dispute board.[10] These events need to be looked at in the context of the evolution of trade unionism on the island, exclusive of the fishing industry, which was a special case. What industry there was was concentrated in St John's, as were the trade unions. There were the bleak, inescapable facts of an overabundance of labour keeping wages down and a cost of living that was 30 to 40 per cent higher than in Canada.[11] The trade unions lacked clout and discipline, and they were often ephemeral.[12]

TRADE UNIONISM IN NEWFOUNDLAND

In the 1930s, there was a renaissance of trade unionism, concentrated chiefly in St John's, with new unions springing up.[13] In 1937, to foster trade unionism, all the unions united to form the Newfoundland Trades and Labour Council, renamed the Newfoundland Federation of Labour in 1939. Not knowing how to handle this unexpected turn of events, the Commission of Government requested the services of an expert on labour relations from the UK government.[14] The latter responded by sending T.K. Liddell, a conciliation officer. He submitted his report in 1940.[15] His unflattering observations on almost every aspect of Newfoundland society had to be toned down or deleted before the report could be released.[16] Liddell's comments on the labour scene at that time in Newfoundland are worth noting since they could be a preview of what was to take place at the Corporation a year later. He was scathing in his portrayal of the Newfoundland workforce: "When it comes to work which requires real efficiency and which is subject to discipline, their lack of education and training is at once obvious ... [They] are easily led into trouble. They are quite willing and anxious to follow any person who will promise them something better than they have." The state of trade unionism was equally dismal: "The majority [of its members], have yet to learn what trade unionism is ... its claims, its rights, its duties, and its obligations." Union officials needed to realize that signing an agreement implied an intention to carry it out and an ability and authority to persuade the membership to abide by its terms. The trade unions, being small, had no funds to hire officials skilled in organizing workers and teaching negotiating skills. In fairness, what Liddell was describing was the birth pangs of trade unionism in Newfoundland, which were no different from what had occurred in England in the earlier phase of industrialization.[17] Liddell's strictures have to be qualified with respect to ASARCO. There, the Buchan's Workmen and Protective Union was affiliated with the American Federation of Labour – an unusual circumstance in Newfoundland in those days.[18] This may account for the union's sophisticated conduct. Wabana was a different story.[19] At Wabana, a union was formed in 1923, only to fold or to become inactive after a few years; management charged that workers were unreliable and ill-disciplined, and workers charged that management was hostile to trade unionism. Only in the 1940s was an effective union established.[20]

Trade unionism did not arrive at St Lawrence until August 1939, with the organization of workers at the Corporation and the American Newfoundland Fluorspar Co. into the St Lawrence Miners and Labourers Protective Union, affiliated with the Newfoundland Federation of Labour, but not registered with the Registry of Trade Unions.[21] In December 1939, without giving notice, the stevedores at the Corporation went on strike, demanding an increase in pay.[22] The ship sailed away with only half its cargo. In January 1940, Seibert increased the basic wage – the first increase since the outbreak of war – to 25.7 cents per hour.[23] Presumably, this gesture ended the strike. However, because of this strike, but also because of other concerns about the labour climate at the Corporation, the commissioner of Public Utilities, W.W. Woods, requested a local justice of the peace, N. Short, to look into the Corporation.[24] In his report, Short commented on the acrimonious relations between the union president, Patrick J. Aylward, and the Corporation's mine manager, Donald A. Poynter. Poynter had refused to recognize the union until it came up with an organizational plan indicating whom it represented and providing evidence of the executive's capability to control the membership and to enforce the fulfillment of commitments. The wildcat strike in December, for which the union executive disclaimed responsibility, Short concluded, pointed to the inability of the union's executive to control the membership and the latter's lack of discipline. Finally, the union would have to learn to negotiate in a reasonable way and to seek advice on how to do so. Nevertheless, labour relations continued to deteriorate. In August 1940, the union and the Corporation reached an agreement that would run for one year.[25] Meanwhile, a feud had developed between the union's president, Patrick J. Aylward, and treasurer, Aloysius Turpin, paralyzing its effectiveness.[26] Claiming that the president was too compliant to Poytner, the rank and file, led by Turpin, seceded and formed a new union with Turpin as president. The new union – The St Lawrence Workers Protective Union (SLWPU) – was registered with the Registry of Trade Unions but was not an affiliate of the Newfoundland Federation of Labour.[27] The membership of the new union grew rapidly and came to include the workforces of both the Corporation and Newfluor.[28]

STRIKES AT THE CORPORATION

While the management of Newfluor enjoyed good relations with the new union, the same could not be said of the Corporation.[29] One bone of contention was Poynter's insistence that the agreement reached with the previous union was still binding until its expiration in August 1941.[30] The failure of Poynter and the union to establish a workable mechanism for resolving contentious issues was to have repercussions. In March 1941, the first of a rapid succession of five strikes occurred. On St Patrick's Day (17 March), workers at both companies took the day off without prior notice or agreement with management. Those at the Corporation were suspended. Another strike occurred on 17 April and lasted until 5 May, precipitated by the hiring of a truck loader who was not a native of St Lawrence. According to Poynter, the people of St Lawrence considered the fluorspar deposits their property to be worked only as they saw fit.[31] In the months that followed, the conflict moved to more substantive issues, such as the recognition of the union as the sole bargaining agent for the workforce and closed shop. The first demand Poynter had no trouble accepting, but the second and other matters would have to be negotiated. This was unacceptable to the union, which turned to the commissioner of Public Utilities and reiterated its request for a board of inquiry.[32] Woods replied that this was unnecessary but offered to set up "a Board of Conciliation."[33] Seibert then offered to require that all new workers would have to join the SLWPU, but refused to compel present workers to join the union or face discharge.[34] In return, he demanded a no strike commitment. In the autumn, the union added a new demand to its list – a wage increase to 45.5 cents per hour.[35] Since the outbreak of war, Seibert had granted several wage increases – the latest on 22 July 1941 to 32.6 cents per hour, the same wage as Newfluor.[36] In October, the union tabled another grievance – payment on time.[37] Since 1940, the Corporation had adopted a system of paying wages in cash. Sometimes there were delays in obtaining sufficient bills and coins of the right denominations from the nearby bank, which handled the Corporation's bankroll.[38] Now the union said such delays would no longer be tolerated, and when such a delay occurred on 15 October, they went on strike. On 1 November, in a letter to Woods, the union's solicitor, P.J. Lewis formulated the union's demands.[39] They wanted recognition of the union as the sole bargaining agent, closed shop, acceptance of the most recent wage increase demand, and payment on time. Lewis went on to accuse Woods of being unreasonable and insisted that the union's demands be referred to an arbitration tribunal or a trade dispute board. The Corporation accepted all these demands except for the closed shop.[40] However, what hurried things along were the warnings from the high commissioner for Canada and the plant manager of DOSCO that fluorspar stocks at the steel mills in Sydney were running perilously low. There was no alternative source of fluorspar, and replenishment was urgently needed before the winter closed navigation.[41] Woods now had no choice but to intervene and get the Corporation back into production. On 10 November, he indicated his intention of constituting a trade dispute board, the terms of reference to include consideration of the union's demands.[42] Turpin, the union president was notified, and he, in turn, called on the men to return to work.[43] They refused until they had received written notification of the appointment of the board and its terms of reference. Even when this was done, some still refused. The last holdouts returned to work on 5 December, after Seibert begrudgingly authorized a temporary pay increase to 45.5 cents per hour, pending the decision of the trade dispute board.[44]

THE TRADE DISPUTE BOARD

The membership of the trade dispute board and its terms of reference were announced on 6 December.[45] The chairman, A.M. Fraser, was a professor of history, economics, and political science at Memorial College.[46] The other two members were W.J. Walsh and Thomas Lefeuvre.[47] The former had been a minister of agriculture in the previous regime and was appointed to represent the union. The latter was a St John's businessman. Newfluor signified its intention to implement the trade dispute board's recommendations.[48]

Before tackling the contentious issues of its mandate, the trade dispute board visited St Lawrence early in 1942 and toured the Corporation's properties.[49] It wrote in glowing terms of what it saw: "We were genuinely impressed by what the Corporation had been able to achieve in a few short years ... This ... was possible only because the Corporation ... enjoyed the close cooperation of the employees and of the merchants of St Lawrence ... The crude [mining] methods of the earlier days are giving place to more established methods." The board's report cited the introduction of the Edison safety lamp (the earliest in Newfoundland), which replaced carbide lamps, as "evidence of the enterprise of the Corporation and of its desire to promote the safety of its workers." As a result, the accident rate was below the average for this type of mine. The hoist house, machine shop, lunch house, washrooms, heating and drying rooms were visited and pronounced to be "modern, efficiently operated and comfortable."

The trade dispute board's somewhat effusive observations on conditions of work at the Corporation appear to be at variance with the comments H.M. Mosdell, secretary in the Department of Public Health and Welfare, had made a year earlier and with Rennie Slaney's account written two decades later.[50] However, it would seem that with improvement in its financial circumstances, the Corporation now had funds to spend on making the work environment more acceptable. This was becoming apparent as early as 1942, according to Slaney. It would stand to reason with another mining company in the town, one with funds to spend on workers comforts, that the Corporation would be pushed into trying to match conditions.

The board was not so enthusiastic about the underground miners. None had had any previous mining experience or pre-employment instruction from experienced miners. It was amazing how well they performed in spite of these handicaps – a tribute, said the board, to their quickness in learning. Yet, this did not alter the fact that their mining knowledge needed to be increased by hiring experienced miners and sending foremen away to modern, well-equipped, and well-run mines for training and instruction. The same lack of knowledge and training was also observed in surface workers. Lack of training was bound to have an effect on output. (Earlier, Poynter had told Short that his men were less productive than those at ASARCO and Wabana.[51]) But, apart from lack of knowledge and poor training, there may have been other reasons for their alleged shortcomings. Fraser cited an incident when unionized stevedores refused to load a boat. It was done in record time by non-unionized labourers.

The board members quickly became aware of the bad feelings between Poynter and the union.[52] In view of the more tranquil workplace atmosphere at Newfluor, the thought naturally arose that some of the blame for this state of affairs had to rest on Poynter. The union accused Poynter of not always being true to his word.[53] Turpin, the union president, claimed that on one occasion Poynter had broken an oral agreement, but when pressed by Fraser to be more specific, Turpin was obliged to withdraw some of his more sweeping charges. Fraser argued that Poynter was well-able to clear himself from "the thoughtless charges made against him by the union." It should be born in mind that the improvement in the Corporation's operations coincided with Poynter's years as manager. He remained in charge of the Corporation until it closed. In the years to come, relations between him and the union seem to have been smoother.

During its tour of the Corporation, because of complaints of poor air quality and dustiness, the trade dispute board checked to see if these charges were justified. No foul air or dustiness was found, so the board claimed, but neither the Iron Springs nor the Blue Beach mines were back in full operations since the partial flooding during the last strike. Furthermore, A.E. Cave, an inspector from the Ontario Department of Mines who had inspected the Corporation in 1939, had made no reference to dustiness. Still, the trade dispute board recommended dry jackhammers only be used if there was plenty of groundwater.

MEDICAL SERVICES AT ST LAWRENCE

The trade dispute board took a look at the history of medical services at St Lawrence.[54] In 1933, there had been none. The nearest community hospital, called the Bankers, was at Grand Bank on the other side of the Burin Peninsula. In 1936, a government-operated cottage hospital had been built at Burin on the same side of the Peninsula as St Lawrence, about twenty kilometres away by boat.[55] Anyone from St Lawrence who required medical attention had to go or be taken to one of these hospitals. The secretary of the Department of Public Health and Welfare, Mosdell had complained of his inability to organize nursing services at St Lawrence due to the "absolute lack of cooperation" of one (unspecified) of the two mining companies.[56] A local medical board had existed for some time. It had already started constructing a clinic, which it was hoped would be an inducement for a physician to settle in St Lawrence.

The union, concerned about the risk for developing silicosis, had been pushing for periodic chest x-rays on its underground members for some time. The trade dispute board refused to make this a recommendation claiming that "we are not competent to pronounce as to the necessity [of x-rays] ... but we desire to place the union's wishes on record as they were expressed very strongly."[57] This is an amazing statement. Silicosis had long been a known scourge of miners. Its clinical and pathological features had been described before World War I. In the interwar years, chest x-rays had become the recognized procedure for early detection of the disease.[58]

CLOSED SHOP

The trade dispute board had a number of contentious issues on which it had to provide guidance to the government.[59] The first was closed shop.[60] The board stated that this would be justified if the SLWPU was under wise and competent leadership, capable of negotiating, and if the membership was alive to the responsibilities of trade unionism. But the recent strikes at the Corporation, so the board claimed, showed that these conditions were not present. Strikes were often called for trivial reasons without authorization or a general vote as prescribed by the constitution. The inability of the union executive to control its membership was demonstrated by the latter's failure to return to work in the previous December when they were told to do so by the union president. Even more reprehensible in the eyes of the board was the walking off the job of essential (maintenance) workers on one occasion, in spite of instructions to the contrary from the union executive. The board therefore rejected the demand for unconditional closed shop, citing as a reason, that if the union was to have a right to strike, management had to have the right to hire non-union workers to maintain essential services. Seibert had agreed to conditional closed shop, however, provided that non-union members be kept on, that there be a no strike clause, and that certain exemptions be made for maintenance service workers, such as firemen. At a general meeting, the membership accepted these terms as well as an arbitration mechanism to run for the duration of the war. The board endorsed this agreement and further agreed that the union was to be the sole bargaining agent for workers.

WAGES

The second controversial issue that required settlement was the matter of wages.[61] There were two aspects to consider: the amount and delays in payment. With respect to the amount, war-time legislation had frozen them at pre-war levels but permitted increases tied to rises in the cost of living. The Corporation's pre-war wage was 23.3 cents per hour. It admitted this was low but claimed that it could not pay higher. The board accepted its word. When it came to estimating what should be a just wage at the time of the trade dispute board, the board's recommendation would have to be consistent with war-time legislation governing wages. It therefore recommended that the average pre-war basic wage should have been 32.6 cents per hour, the same estimate that Newfluor had been using to calculate increases in its war-time wages. The board noted that thirteen cents per hour should be added to account for an increase in the cost of living since 1939, which brought the average total to 45.6 cents per hour, almost the same as the Corporation was now paying temporarily under protest. But could the Corporation go on paying at this rate? The Corporation's commitment to modernization and expansion of its holdings worked in its favour. The salaries of the management were modest. It had never paid dividends but instead had ploughed back profits into improving its operations. The board was of the opinion it could just afford its recommended wage package, provided there were no more strikes. Looking to the future, to monitor changes in the cost of living, the board recommended the appointment of a local committee with equal representation from the union and management and an independent chairman chosen by them. Periodic wage adjustments should occur when the rise in cost of living reached 5 per cent higher than at the previous wage increase. As mining was speculative and profits unpredictable, the Corporation might find itself unable to increase wages in line with rises in cost of living. Therefore, the Corporation should be permitted to appeal to a trade dispute board for exemption from having to do so. The board also examined the the practice of giving a production bonus of five cents for every bucket in excess of the thirty buckets per shift, which had been instituted in August 1941, as workers complained it caused neglect of safety. The board declared that there was no evidence of negligence and urged the continuation of the bonus.[62]

Delays in payment were the last of the workers' financial grievances to be addressed by the board. Delays had occurred more than once, and, on one occasion, delayed payment was due to insufficient funds of the right denominations in the bank. However, the board asserted that with increasing sales and the reversion to payment by cheque, delayed payment should be a thing of the past.

Some of the board's other recommendations, such as time-and-one-half for overtime, cost-of-living bonus, and the union as the sole bargaining agent, were similar to those of a trade dispute board that was constituted to settle a strike at ASARCO about the same time as the St Lawrence trade dispute board. The terms of this settlement, which came to be known as the Dunfield settlement, named after its chairman, Mr Justice (later Sir) Brian Dunfield, became the model for settling war-time labour disputes.[63]

The government took no action to enforce the recommendations of the St Lawrence trade dispute board, leaving it up to the parties concerned to implement them by mutual agreement. In the ensuing months, the recommendations were implemented, first at Newfluor and then at the Corporation.[64] During the remaining war years, the Corporation was able to give wage hikes in step with increases in the cost of living.[65]

In the years that followed the trade dispute board, working conditions continued to improve at the Corporation.[66] Modern equipment was installed – wet instead of dry drills and mucking machines. The three eight-hour shifts were replaced by two eight-hour shifts, which permitted the mines to be blown out by compressed air between shifts.[67] In 1944, a new contract was reached between the Corporation and the union whereby the Corporation would be responsible for providing toilets, drying rooms, and washrooms as well as a mess.[68] Rubber boots, gloves, and helmets were to be sold to workers at retail prices. An arbitration mechanism was instituted, pursuant to war-time regulations, requiring referral of a dispute before a strike or lockout to a Mines Advisory Committee with equal representation from the union and management. Should the committee be unable to resolve the dispute, it was to be referred to the labour relations officer for a ruling that would be binding on both parties. (This official, A.J. Walsh, had been appointed in 1942, as a functionary in the Department of Public Utilities with oversight of industrial activities.[69]) Throughout the remaining war years, labour relations at both companies followed a smooth course, apart from a few short-lasting work stoppages.[70] However, for the Corporation there was an ominous cloud on the horizon. By 1944, Mexican fluorspar was being delivered to US markets at five dollars per ton less than that from St Lawrence.[71] Unless the Corporation could become competitive its future was bleak. However, the political scene was about to change once again in Newfoundland.

THE LAST DAYS OF COMMISSION OF GOVERNMENT

When World War II ended, Newfoundland was still under Commission of Government. With the island now financially solvent thanks to the war, there was no need for the Commission's continuation. It was obvious its days were numbered. For the present, the order of the day to the civil service was to mark time, leaving it up to the new regime to introduce what measures would be required to speed up the process of catching up with mainland North America. Perhaps in no department was this philosophy more obvious than in the Department of Public Utilities. There, C.K. Howse, by now in charge of mines, was told his budget would be approved only if it was the same as the previous year.[72]

THE MINING INDUSTRY COMMITTEE

In 1946, the UK government convened a national convention in St John's to advise on the political future of the island. As part of its work, the convention constituted committees to report on aspects of the social and economic features of the island. One of these committees was the Mining Industry Committee. The committee's report emphasized the intractable handicap facing any industrial enterprise in Newfoundland – the high cost of production.[73] For the Corporation, the cost of production was estimated to be 40 per cent more than for its competitors. The reasons were the high cost of the goods it imported, custom duties, and wage increases to keep pace with war-time rises in the cost-of-living index (set at 100 in 1939 and reaching 165 by 1945). For the fluorspar mines at St Lawrence there was the added burden of paying for the large amounts of electricity required to keep the pumps running to empty the mines of the constantly incoming floods of water. Hydroelectric power cost 60 per cent more than in Canada. The report also noted the different war-time courses the two companies had followed. The Corporation, following the loan it had received from the US government, had shipped increasing amounts of its product to the United States; in 1944, 75 per cent of its output went to the United States. By 1945, that number dropped to 48 per cent and stayed around that percentage for the next few years. The rest was sold as metallurgical grade to the steel mills of eastern Canada.[74]

THE ST LAWRENCE FLUORSPAR INC

To further process and distribute his fluorspar imports to the United States, in 1946, Seibert incorporated his plant at Wilmington, Delaware as the St Lawrence Fluorspar Inc.[75] But after the war, the financial circumstances of the Corporation took a turn for the worse. Demand for fluorspar dropped.[76] Creditors were pressing for repayment of loans while banks would not provide further credit.[77] The one bright spot was that, when it came to wages, the union was accommodating.[78] On the other hand, Newfluor, having stockpiled 30,000 tons of fluorspar by 1945 (twice Alcan's yearly requirements), suspended operations and allowed the Director mine to flood.[79] (It reopened in 1948.[80])

MINES INSPECTIONS

In the post-war years, Commission of Government continued its pre-war practice of periodically calling in a mining engineer to inspect the premises and pits of the four mining companies and the one limestone quarry on the island and to make recommendations. In this capacity, Professor A.V. Corlett of Queen's University, Ontario, paid three visits, each in the late summer – the first in the last year of the old regime (1948), and the other two in the first two years of the new (1949). Corlett's inspections focused entirely on safety and workplace hazards. Newfoundland mines, he noted, compared favourably with Canada with respect to accident rates.[81] For the four mining companies on the island, the rates (frequency per 1,000 workers per year) were 131 for each of the two fluorspar companies, 155 for Wabana, and thirty-nine for ASARCO. (Sixty was considered to be an acceptable rate.) Not surprisingly, Corlett reserved his highest praise for ASARCO and his most astringent comments for Wabana. He described Newfluor as "carefully run, doing its best in safety operations and capital not a problem." The Corporation he characterized as "having grown up the hard way ... but needs more improvement ... The accident rate is better than would be expected which is due to the attitude of management towards the crew rather than the physical plant." The Mining Industry Committee of the National Convention had cited the accident record of the Corporation as "excellent" and, for its years of operation, only one-third of what would be expected in equivalent types of mining.[82]

By the following year, accident rates at both fluorspar mines had doubled.[83] But, as Corlett explained, this might be more apparent than real due to changing the definition of a lost-time accident from six to three days off work under the new Workmen's Compensation Act of 1948. Still, he pointed out, it would be unwise to ascribe the increase entirely to a change in definition. Assuming there had been some increase in the accident rates, what were the reasons? Corlett suggested that, at the Corporation, the increased accident rate was not due to a widespread violation of ordinary precautions but "the result of [a] lack of appreciation of hazards." What were needed, he wrote, were strict regulations, inspection, and compliance. On his first visit he had been asked to recommend safety measures that could be codified and incorporated into a revision of mining regulations.[84] The authorized Code of Mining Practices dated back to 1916; it was out of date and had many flaws. An operator could appeal against an inspector's decision to an arbitration board. The code left it up to an operator to draw up his own safety rules within limits set by the act and subject to certification by the government engineer. In Newfoundland, operators used the *Handbook of Regulations* issued by the Ontario Department of Mines, but only as a guide. At times, the handbook was ignored. In general, Corlett wrote, there was a nonchalant attitude to safety and inadequate indoctrination into safe practices, especially at the Corporation.[85]

This careless attitude towards safety was a recurrent complaint levelled at the St Lawrence miners. Perhaps one explanation is to be found in Price Fishback and Shawn Kantor's observation that, following the introduction of workmen's compensation into industries, while rates for non-fatal accidents rise, they do not for fatal ones.[86] He ascribed this pattern of behaviour to the capability of workers to distinguish between conduct that can have serious outcomes from that which does not. Before workmen's compensation, it was to the financial advantage of workers to avoid all lost-time accidents, an incentive that the coming of workmen's compensation diminished. He added that this carelessness was particularly true of the mining industry because of the lack of supervision and the freedom miners enjoyed in the performance of their tasks. Corlett had looked at other Canadian provincial mining codes, but thought none were applicable to Newfoundland.[87] He wrote, "Under the best of management there is a tendency to go no further than the law requires." He concluded that it was better to leave it up to the new regime to promulgate an up-to-date code of operations. Corlett referred to the rock bursts, which were a plague in the fluorspar mines, and warned that they could be a greater problem in the future.[88] He had nothing to say about early warning signs of impending rock bursts.

With only four mining operations and a limestone quarry on the island, a good case could be made for continuing the present system of periodic inspection of the mines by a qualified mining engineer brought in from the outside. Instead, against Corlett's advice, the government chose to advertise (unsuccessfully) for a highly qualified mining engineer.[89]

Corlett's reports demonstrate the limitations of mine inspections in those times. The focus was entirely on safety. There was no reference to dust conditions nor to measurements of ambient dust. Only in his last report in 1950 does he refer to concerns about silicosis, raised by the admission of a miner to the St John's Sanatorium with a provisional diagnosis, which later was not confirmed, and a case of tuberculosis with possible coexistent silicosis.[90] Corlett was of the opinion that because of the wetness of the mines silicosis would not be a problem. This was to ignore the mill, which was not deluged daily with water like the mine pits and, therefore, was where the millers might not be protected from developing silicosis.

With the end of Commission of Government the two mining companies went their separate ways – one providing exclusively for an aluminum smelter, and the other providing mainly to an American market.

3

Newfluor in the 1950s

On 31 March 1949, following a referendum, Newfoundland became Canada's tenth province. The premier of the new government, J.R. Smallwood, had been active in organizing trade unions in his earlier days. During the National Convention, he had advocated updating the island's labour laws. One of his first actions after assuming office was to appoint a labour advisory board to examine labour relations in the other provinces and to submit recommendations for bringing Newfoundland's labour laws in line with the rest of the country.[1] With the exceptions of the chairman and secretary, all the other members of the board had been active in the labour movement of the island. The board submitted drafts for legislation regulating trade unionism, Workmen's Compensation (WC), and Workplace Health and Safety – the last two now combined into a single organization, Workplace Health, Safety and Compensation (WHSC). These drafts in turn became the basis for laws. By 1951, Newfoundland was said to have one of the most enlightened corpus of labour laws in the Western world.[2] The new Workmen's Compensation Act, which came into force in 1951, was modeled on that of Nova Scotia.[3] Previous workmen's compensation laws had followed the United Kingdom system whereby industries insured themselves with private carriers for legally stipulated benefits for workplace accidents and illnesses. Now Newfoundland switched to the Canadian system, by which employers pay an annual premium to a Workmen's Compensation Board constituted by the provincial government and responsible for paying compensation for workplace accidents and illnesses.

NEWFLUOR

In 1948, after a three-year shutdown, Newfluor started up again.[4] Operations expanded during the next decade.[5] In 1951, a heavy media separation plant was constructed, which permitted the mining of low-grade ore and raised the calcium fluoride (CaF_2) percentage to 70–80%. This product was then shipped to Alcan's aluminum smelter at Arvida, Quebec for further refining to the level of acid grade (92%) – the grade required for aluminum smelting.[6] Looking back, it seems wasteful for Newfluor never to have brought its fluorspar up to the required acid grade level at St Lawrence. But to have done so, according to an informant, would have required closure of the Arvida flotation mill, an action he suspected the Arvida Smelter Union would not countenance.[7] Expansion of the Director mine was completed by 1957.[8] Output fluctuated from year to year, presumably governed by the needs of the Arvida Smelter. (See Appendices C and D.)

ALCAN'S LONG HARBOUR CLAIM

Relations between Alcan and the Newfoundland government were generally cordial during the Smallwood years,[9] apart from an incident that was smoothed over before developing into a full-blown row. The circumstances were these: in 1947, Newfluor staked two claims at Long Harbour, Fortune Bay. During the next few years a considerable amount of money was expended prospecting for minerals, and what appeared to be a promising vein of fluorspar had been discovered. In 1952, like a bolt out of the blue, Newfluor was told that its prospecting activities were illegal because the claims were on land where the mineral rights had been reserved to the Crown (the provincial government) under the Lands Act of 1950.[10] No one had thought of telling Newfluor until this abrupt notification. Alcan's president at the time was R.E. Powell – not a man to tackle lightly, according to those who had dealings with him. He was outraged. The reason the government wanted to eject Newfluor from its Long Harbour claims was that the Newfoundland and Labrador Corporation (NALCO), the Crown corporation set up by Smallwood to exploit the natural resources of the province, wanted to sink a shaft of its own on the claim occupied by Newfluor and to get into the fluorspar business. Alfred Valdmanis, the president of NALCO, was eager to get going, but Minister of Mines F.W. Rowe and his deputy minister, C.K. Howse, were sceptical of his chances of success. They told Smallwood of their reservations about Valdmanis's project and that in their opinion Newfluor had a "moral right" to the Long Harbour claims. Powell threatened to suspend all further prospecting operations unless he was left in possession of the Long Harbour claims. Smallwood bowed to these pressures, and Newfluor was left in possession of the claims.[11] But, whatever the reason, the Long Harbour deposits were never developed.

MINERS' HEALTH PROBLEMS

If the 1950s were a period of expansion and prosperity for Newfluor, they were also a time of growing unease about the health of the miners.[12] These concerns went back to the mid-1940s when the workers and their families began to become suspicious that underground mining in the St Lawrence fluorspar mines was not a healthy occupation. Those who engaged in it were not only coming down with tuberculosis at an alarming rate and were more likely to die from it than others, but the miners also seemed to be generally unhealthy for some unexplained reason.[13]

At first, these rumours were confined to the employees and their families, but by 1949, Dr J.J. Pepper, the resident physician in St Lawrence,[14] began to have his own concerns about the health of the miners.[15] More to the point, he was stirred into action to find out the reason for the miners' vague ill health, other than tuberculosis.[16] His lobbying initiated a series of investigations that were to have a surprising outcome ten years later. But in 1949, he wondered whether silicosis might be the other explanation for the miners' mysterious ill health.

At the time of the trade dispute board in 1942, the union had pressed Fraser, chairman of the board, to recommend that chest x-ray surveillance be made mandatory for the miners.[17] His refusal to do so, even though he did incorporate the union's request in his report, meant that regular x-ray surveillance of the miners was delayed until 1947, and, from then until 1954, it was conducted in a haphazard fashion. In contrast, by 1945, ASARCO and Wabana both had installed chest x-ray machines at their operations.[18] In 1947, the MV *Christmas Seal*, a boat owned and operated by the Newfoundland

Tuberculosis Association for conducting tuberculosis surveys in the outports, paid its first visit to St Lawrence. It returned in 1950 and 1954.[19] Free chest x-rays were offered to everyone in the community. The aim was to detect tuberculosis. If other chest diseases were present, such as silicosis or cancer of the lung, they would also of course show up in the x-ray. The 1947 visit of the *Christmas Seal* revealed three cases of suspicious tuberculosis.[20] Corlett reported that by 1950 there were several cases of suspicious silicosis and silico-tuberculosis.[21] None of these suspicious cases were confirmed. Thus, when Pepper began his campaign for a thorough investigation for silicosis in the mines, he had little evidence to back up his plea. Indeed, the management of both companies and Professor A.V. Corlett discounted his fears, arguing the mines were too damp to pose a hazard for silicosis.[22] In fact, it was not until 1952 that the first indubitable diagnosis of silicosis was made.[23]

HEALTH SURVEYS

In 1950, shortly after Newfoundland became a province of Canada, the federal Department of National Health and Welfare (NHW), in conjunction with the Newfoundland Department of Health, carried out a health survey of industries in the new province.[24] In its final report, it recommended comprehensive studies on the health status of employees in all industries in Newfoundland. The team spent three weeks in St Lawrence. During its stay, Pepper expressed his fears to the visitors that besides tuberculosis, silicosis might also be another reason for the vague ill health of the workforce, but so far he had not been able to confirm this.[25] Shortly afterwards, on his own volition and apparently without consulting Newfluor, the Corporation, or the provincial Department of Health, he wrote to National Health and Welfare requesting "a proper survey and dust count."[26] In reply, he was told it was up to the Newfoundland Department of Health to issue the invitation. (In Canada, health is a provincial responsibility.) Pepper was not popular with the management of the two mining companies. His well-intentioned interventions on behalf of the miners were viewed as meddlesome.[27] What Pepper was pushing for was an examination of every miner, including a chest x-ray, and quantitative measurements of the ambient dust underground.[28] Quantitative dust measurements had not been performed previously in the St Lawrence fluorspar mines, but had been done in the United States since the mid-1920s, using the midget impinger.[29] Unfortunately, this instrument did not give consistent reproducible estimates.[30] Still, it was a step in the right direction towards establishing safe dust exposure limits, designated Threshold Limiting Values (TLV). In 1946, the American Conference of Governmental Industrial Hygienists (ACGIH) issued the first of its recommended safe exposure limits, TLV, for mine dust.[31]

As mentioned earlier, Pepper's lobbying set in motion a series of government actions. An understanding of these events requires a knowledge of the functioning of occupational health and safety (OHS) in the province in those days. During the Smallwood years, responsibility for OHS was shared between the Departments of Health, and of Mines, Agriculture, and Resources (referred to as Mines henceforth). The former was responsible for health surveillance and the latter for health and safety inspection as well as monitoring the engineering controls required to safeguard workers' health and safety. The system worked reasonably harmoniously and effectively, apart from the occasional charge by one department accusing the other of encroaching on its turf.[32] The major flaw in this scheme was that it was reactive, rather than preventative.

SILICOSIS

To return to Pepper's prodding for a full-scale investigation, the provincial government must have taken his concerns seriously because, in June 1953, dust measurements (TLVs) were underway using the midget impinger.[33] Most sites were within what was regarded as safe limits. This did not exclude the possibility that, in the past, TLVs might have been dangerously high, in which case, the authorities were told to prepare to deal with a flood of cases of silicosis. To make sure that dust levels would be kept safe in the future, a meeting was convened by the Departments of Health and Mines with experts on silicosis from Ontario. At this meeting, officials decided to develop a code of practice for the prevention of silicosis.[34] The code was enacted into law in 1959. In fact, those who claimed the mines were too damp for silicosis to develop were proven right. The flood of cases of silicosis never transpired. An epidemiological survey of the mines published in 1964, following strict scientific diagnostic criteria, identified five definite and five suspicious cases of silicosis[35] – a prevalence considerably lower than in the UK coal mines at that time.[36]

TUBERCULOSIS

When it came to tuberculosis the story was different. Morbidity and mortality statistics collected some years later confirmed the community's impression that the miners were prone to developing tuberculosis as shown in the rates for the years 1945–55.[37] There are two explanations for the miners' susceptibility to tuberculosis. Working in closed spaces at close quarters would be an ideal setting for spreading the disease from open (infective) cases. Then there is the well-known synergistic relationship between silicosis and tuberculosis, functioning at amounts of silica inhalation insufficient to show up in radiographs as nodules.[38]

NATIONAL HEALTH AND WELFARE SURVEY

Thus, by the mid-1950s, with only a few cases of silicosis, the data on tuberculosis not yet in, and the dust measurements no cause for alarm, Dr Leonard Miller, the provincial deputy minister of health was in a quandary about what to do next to get to the bottom of the reason for the miners' apparent ill health.[39] At this point, the Department of Mines came to the conclusion that it was incumbent to have the accuracy of its dust measurements verified. Accordingly, the department wrote to National Health and Welfare requesting a dust survey but was told that the invitation would have to come from the Department of Health.[40] Sometime late in 1955 or early 1956, the Department of Health issued the necessary request, which was subsequently accepted.[41] By the autumn of 1956, J.P. Windish and H.P. Anderson, industrial hygienists from National Health and Welfare, were conducting dust measurements in the mines. The measurements were completed six months later. They found dust measurements above permissible levels only at some stations in the mill and at a few mine sites.[42] In other words, dust levels were safe except at a few locations. These results were communicated to the Departments of Health and Mines. Miller was no further ahead in solving the problem of the sickly workforce.

FIRST CASES OF LUNG CANCER

To get the epidemiological part of the study underway, Dr A.T. deVilliers, a consultant in the Occupational Health Division of

National Health and Welfare, which was responsible for conducting the survey, visited the island in December 1956 and started collecting data. During his visit, he was told a startling item of news by Dr C.J. Walsh, a local physician; during the previous two years there had been six deaths from cancer of the lung among the fluorspar miners. DeVilliers immediately passed on this disturbing news to Dr F. deN. Brent, the chief medical officer at Alcan.[43]

C.J. Walsh had come to St Lawrence in 1954 as the physician in charge of the newly opened cottage hospital.[44] Within the first year of his arrival, Walsh noticed a high death rate from cancer of the lung among the miners. He is reported as having informed Workmen's Compensation and the Department of Health of this disquieting development, although it is not recorded when.[45]

At that time, no one would have faulted Walsh for ascribing this cluster of cases of lung cancer to smoking since a high percentage of the St Lawrence miners were heavy smokers.[46] Furthermore, 1954 was the year R. Doll and A.B. Hill published their landmark paper, "The Mortality of Doctors in Relation to Their Smoking Habits," which did so much to convince the English-speaking medical profession of a link between smoking and cancer of the lung.[47] Walsh showed astuteness in his observation that, at the cottage hospitals where he had practiced previously, cancer of the lung was "very rare" but smoking habits were similar to St Lawrence.[48] It is important to realize that at this stage Walsh's observation could only be regarded as a suspicion. As with other examples of occupational and lifestyle cancers, enough cases to permit statistical analysis would have to be accumulated before the scientific and medical communities would be convinced of the validity of his hypothesis. This was to take almost ten years.[49] Nevertheless, both Alcan and National Health and Welfare seem to have taken Walsh's observation seriously and realized that the mining companies could be facing a potentially catastrophic situation – an epidemic of cancer of the lung.

BRENT'S PROPOSALS

Upon receipt of deVilliers's letter, Brent applied to Alcan's head office for permission to visit St Lawrence to judge for himself if what Walsh said was true and, if so, what should be done.[50] Permission being granted, he visited St Lawrence, 23–29 October 1957. He submitted his report to Alcan's head office in February 1958.[51] His report was a milestone in the detection of the hazard responsible for the cancer of the lung affecting the miners. In it, he listed the measures that would have to be followed to confirm Walsh's hypothesis: first, a prospective study that collected enough data on the miners' causes of death to permit statistical analysis would need to be conducted; second, dust measurements would need to be taken; third, the mine dust would need to be tested for carcinogenicity in animals; and finally, the air in the mines would need to be checked for radon and radon daughters.[52] (Radon daughters had been shown to be the cause of alpha radiation-induced lung cancer in 1951, the same year in which a reliable instrument for their measurement had been invented.)[53]

Once it became clear that an epidemic of lung cancer was developing in the St Lawrence fluorspar mines, Alcan became concerned at the public relations aspect of the calamity. Brent wrote to Rupert Wiseman, the plant manager, explaining that while the company did not wish to be "bleeding hearts," it would be unfortunate if it were to appear "callous." Since it would be up to Wiseman to "carry the ball," Brent wished to be kept informed about how Wiseman proposed to handle the inevitable publicity that would arise.[54] Brent's public relations concerns proved unnecessary. Once it was realized that a serious health condition might be developing in the mines, control of events passed out of the hands of Alcan into those of the federal and provincial governments. All that Brent could do was to keep deVilliers (who was still conducting his dust studies) informed of his thinking on what should now be done to find the cause of this new alarming development. This he did with tact, and deVilliers seems to have welcomed his suggestions. The two men and Leonard Miller, the provincial deputy minister of health, worked together tirelessly on getting to the bottom of the problem.

In the months that followed his visit to St Lawrence, Brent kept in close touch with deVilliers and sent him a copy of his report.[55] DeVilliers told Brent that National Health and Welfare was not equipped to check for radon and its progeny in the mines.[56] But Brent did not stop looking for a company competent to undertake the task. He found that a company in Boston, Massachusetts called Arthur D. Little Co. could perform the testing.[57] Alcan appears to have been reluctant to follow Brent's suggestion and do a radon and radon progeny check in the mines. Brent thought Alcan was reluctant because he was not an expert on radiation. It would take more than his word, he realized, to persuade the company to embark on what might be a fruitless and expensive venture.[58]

Brent had no intention of leaving matters at a standstill. He wrote to Miller pointing out that the only way of moving ahead was to initiate an epidemiological study and to check the mining environment for a cancer-inducing agent.[59] He advised Miller to ask National Health and Welfare to undertake these tasks. Acting on this advice, Miller requested National Health and Welfare to expand its studies to include an investigation on the outbreak of cancer of the lung in the miners.[60] National Health and Welfare agreed.

THE CAUSE OF THE LUNG CANCER

At this point, it was assumed that the increased incidence of lung cancer must be related to mining experience of some sort. But if so, what was the agent causing it? Radiation seemed to be excluded because a Geiger counter check of the mines in 1952 "for gamma and beta" radiation had showed "none."[61] (Later Wiseman and J.W. Cameron, the manager and managing director of Newfluor respectively, wondered if the instrument had been faulty.[62] More likely it was not sensitive enough to pick up alpha radiation.) It was therefore understandable that deVilliers and his co-workers in National Health and Welfare should focus their attention on the possible carcinogenic properties of the dust generated by mining. Animal experiments, testing various components of the dust for carcinogenicity, were performed without success.[63] In late 1958, deVilliers turned to Brent's suggestion, proposed months earlier, to check for radon and radon progeny in the air of Director mine, and some preliminary sampling was done.[64] Using a Geiger counter, he found non-hazardous levels of "gamma and beta radiation" originating from the underlying granite.[65] However, it was not until May 1959, following consultation with Miller, that a decision was made to conduct a thorough investigation for radon and radon progeny in Director mine.[66] Radiation measurements were performed by J.P. Windish in November 1959. Using a sensitive scintillometer, high levels of radon and radon progeny were detected in the mine, especially in unused portions and dead ends – areas where the workers were forbidden to enter.[67] Repeat measurements in December confirmed these high readings.[68] Brent later said that his suggestion to check for radon and radon progeny was an example of serendipity.[69] He recalled that in 1953 he had attended a lecture in the United States where the speaker had reported that high levels of radon and radon progeny had been found in the Colorado Plateau uranium mines where alarming numbers of cases of lung cancer were appearing. Fears must have been expressed at the lecture that soon the United States might be facing an epidemic of lung cancer in its uranium mines similar to the one that, for centuries, had plagued the notorious mixed metals and pitchblende (uranium) mines in Schneeberg and Joachimsthal in the Erzgebirge Mountains, which border Germany and the Czech Republic.[70] Such fears were confirmed in 1960.[71] Still, Brent never wavered in his belief that the cancer of the lung at St Lawrence was the result of the interaction of more than one agent.[72] In later years, he wondered if smoking might not be one of these cofactors.[73] He resisted pressure to ascribe the lung cancer solely to smoking as he was urged to do.[74]

On 30 December 1959, in a letter from T.W. Patterson, Chief Occupational Health Division, National Health and Welfare, Miller was officially informed of the results of Windish's radiation survey and Windish's opinion that enhanced ventilation would bring alpha radiation levels to within safe limits.[75] (Radon progeny, the source of alpha radiation has a short half-life; only a few minutes and it is quickly dispersed by adequate ventilation.) Presumably, Windish was drawing on the United States's experience with their uranium mines for making this recommendation.[76] The uncompleted epidemiological study, Patterson said, should be continued to completion, provided funding could be found. The letter was silent on whether the mines should be shut down until the additional ventilation systems had been installed. However, Windish had previously expressed his opinion verbally that the time required to install additional ventilation would be so short that it would not add appreciably to the risk of developing lung cancer.[77]

Newfluor did not wait until Windish had taken his second set of readings and had begun to install additional ventilation machinery in December 1959. The job was completed four months later. Still, Alcan also took care to have Windish's radiation estimates confirmed and retained Arthur D. Little Co., Boston, to conduct an independent radiation survey of Director mine. Using different equipment, Little came up with essentially similar alpha radiation measurements.[78]

On 1 March 1960, Premier J.R. Smallwood convened a meeting. Present, among others, were the provincial ministers of Health and

Mines, officials from Alcan and the union, and the member of House Assembly (MHA) for St Lawrence.[79] The purpose of the meeting was to inform those present of the results of Windish's investigations and his recommendations. Following the meeting, the minister of health, Dr James McGrath released a ten-point public statement summarizing the information released at the meeting.[80] Besides reiterating the elevated cancer of the lung rate among the miners, the high radiation levels in the mines and the remedy – increased ventilation – several other points were mentioned. He reported, dust levels "did not show excessive amounts of silica"; the tuberculosis death rate for St Lawrence in 1952–54 was higher than the average for Newfoundland; and the final point, this was the first time an excessive rate of lung cancer had been found in association with "ordinary mining operations" in the Western hemisphere. Although it yet awaited epidemiological confirmation, the St Lawrence cluster of lung cancers was the first to be reported in non-uranium mines. To put things in perspective, it was not until 1962 that the United States accepted its first case of industrial lung cancer from other than uranium mines, and the UK went on ascribing such cancers to smoking even longer.[81]

THE SOURCE OF THE RADON

Having identified high radiation levels in the mines, the next question was the source of the radon and its progeny. The explanation seems to be this: the fluorspar veins were embedded in granite and rhyolite porphyry.[82] Like all granites the St Lawrence granite contained uranium but twice the usual amounts present in other granites. Radium is one of the decay products of uranium. It decays to radon and then, finally, into radon progeny. Radon and radon progeny are volatile gases. The St Lawrence granite contained many cracks, which were saturated with water because of the heavy rainfall. It was into these water-filled cracks that radon was being released and going into solution. This radiation-contaminated water was then seeping into the mines in groundwater. In the mines, radon was then being released into the surrounding air and breaking down into radon progeny, the cause of the lung cancer. Because of insufficient ventilation, radon and radon progeny were rising to hazardous levels.

The epidemiological study was completed and published in 1964.[83] It confirmed that there was a statistical increase in lung cancer among the miners, most likely due to exposure to high levels of radiation. Whether smoking was a contributory factor was impossible to settle because of insufficient information.

The question naturally arises whether cancer of the lung was one of the explanations for the apparent poor health of the miners in the 1940s.[84] The first proven case had died in 1949 and the next two in 1952. In the years up to 1967, there were two to four cases a year, with a high of eight in 1965.[85] Were there cases before 1949 that had been overlooked? Theoretically, because of the above rates, there could not have been many. Rick Rennie refers to a case in 1944.[86] Furthermore, the average period of exposure to radiation up to diagnosis in the epidemiologic study was 12.5 years with a range of 5.5 to 21.3 years.[87] Exposure to hazardous levels of radiation could only have begun in 1936 when underground mining commenced. Hypothetically, this puts 1941 as the earliest date for the first cases of lung cancer to become apparent. It is highly unlikely few, if any, cases of lung cancer would have been missed in the 1940s for the following reasons. After 1942, a physician had been almost continuously in residence. The chest x-ray machines at the nearby cottage hospitals, Burin and Grand Bank, were available to him. The MV *Christmas Seal*, the tuberculosis screening survey boat, had visited St Lawrence in 1947, 1950, and 1954. Anyone with a "suspicious" chest x-ray would have been referred to the sanatorium in St John's for further investigation. This leaves tuberculosis and chronic bronchitis and its complications, both of which showed increased rates in the mines, as the most likely explanation for the community belief that their miners were an unhealthy lot.[88]

Windish's emphatic assurances that there was now no radiation risk in the mines was based on the then current knowledge about health effects of alpha radiation. In the years to come in the worldwide mining industry, alpha radiation levels were reported to be lowest in the St Lawrence fluorspar mines, adding weight to his optimism.[89] Still, only time would vindicate or refute his assertions. The government and Alcan accepted his assurances. The workers' union was less convinced and quick to publicize breaches of permissible safety levels of alpha radiation. As a result, the public came to share the union's unease. What had been a local concern now became a national one.

4

Newfluor in the 1960s

The public statement of 1 March 1960 did not end the debate about the health and safety of the St Lawrence fluorspar miners. The workers' (both underground and surface) response to Minister of Health Dr James McGrath's public statement was to vote to walk off the job until the new ventilation system had been installed.[1] (Later Alcan was to claim that the union president, A. Turpin, had been informed of Windish's findings and recommendations in his interim report to the minister of health.[2] Management, in retrospect, blamed the walkout on press distortion, a charge the press vehemently denied.[3]) Following the meeting of 1 March, to make sure the union was fully informed of the findings and the remedy, the minister of health sent a copy of Windish's interim report to Turpin. This did not stop the strike, which began in the middle of March.[4]

Windish returned early in April, after the new ventilation machinery had been installed, checked alpha radiation levels, and found them to be in what were deemed to be within safe limits.[5] He informed Turpin in writing of his findings, and in turn Turpin called a meeting of the union membership and told them it was safe to return to work.[6] Sixty of the ninety strikers immediately did so. What to do about the holdouts then became a topic of debate. Alcan argued that unless a worker had an abnormal chest x-ray or suspicious cancer, he should be told to go back to work.[7] Still, understandably, there were workers who now knew they had been exposed to high doses of radiation and who wanted assurances that going back to work would not add to their risk for developing lung cancer. All that Walsh, their physician, could tell them was that it was impossible to be absolutely sure.[8] Eventually, all but sixteen returned to work. They were provided with surface jobs.[9]

Worker discontent regarding the ongoing epidemiological study and questions about the radiation safety of the mines continued to surface throughout the decade that followed, which the press took notice of from time to time.[10] One complaint from the workers was that they were kept in the dark about the results of their physical examination and chest x-rays as participants in deVilliers's yet-to-be-completed epidemiological study.[11] The basis of this charge was deVilliers's custom of only informing participants whose examination and x-rays revealed abnormalities and assuming that participants who did not hear from him would know nothing wrong had been found. Later, deVilliers incurred further criticism for his failure to send a copy of his published epidemiological study to Turpin.[12] It was only in 1965 that Turpin became aware of this paper thanks to a copy he received from Rennie Slaney.[13] DeVilliers even found himself accused of using the miners as "guinea pigs."[14] On the other hand, Dr E. Quinlan, one of the cottage hospital physicians, was singled out for praise for explaining to the miners, in simple terms that they could understand, the nature of the disease they were facing.[15]

COMPENSATION

Once it was apparent that Newfluor was probably facing an epidemic of occupational lung cancer in its workforce, what to do about it became a responsibility of the federal and provincial governments and their advisors. The two most pressing issues facing the authorities were compensation for the victims and their families and whether those who had been exposed to high levels of radiation should be allowed to continue working underground now that radiation levels were considered to be safe. The matter of compensation would require an amendment to the Workmen's Compensation Act, adding cancer of the lung from exposure to radiation to the Schedule of Industrial Diseases qualifying for compensation. Alcan urged Workmen's Compensation Board (WCB) to make this addition to the act.[16] National Health and Welfare (NHW), to which the Newfoundland government looked for advice, hesitated about committing itself to this change because, as deVilliers pointed out, it would be the first addition of its kind in Canada.[17] Brent, a tenacious proponent of the multifactorial etiology of the St Lawrence miners' lung cancer, observed that if compensation was to be based on the proportion of a miner's disease due to occupation, it would be impossible to come up with a formula that could realistically apportion responsibility quantitatively between occupational and non-occupational factors.[18] (This difficulty did not arise with the other diseases then listed in the WCB Schedule of Industrial Diseases, such as silicosis, that were all caused entirely by occupational exposure.[19]) There was also a suspicion that smoking must be a contributing factor to the epidemic of lung cancer at St Lawrence. How this could or even should be factored into a compensation formula was debatable. Finally, deVilliers's definitive study, which conclusively established an increased incidence of lung cancer among the miners, had yet to be completed and published.[20]

In spite of many reasons for reflection, the climate of public opinion forced the government to act. The Workmen's Compensation Act was amended to add "carcinoma or malignant disease arising from radiation" without qualification to the Schedule of Industrial Diseases entitled to compensation, to take effect 5 July 1960.[21] There were two stipulations: the diagnosis had to have been made after 1 April 1951 – the date the first provincial Workmen's Compensation Act came into effect; and the claim had to be made within twelve months from cessation of employment.[22] Alcan argued in vain for an additional stipulation of at least ten years of employment underground.[23]

CONTINUATION TO WORK

If the provincial government was expected to enunciate a policy for dealing with the current group of underground workers who had been exposed to high levels of radiation, it disappointed such hopes. In fairness, at that time, there was no firm scientific evidence to guide it into either transferring all the underground employees to surface positions or letting them continue in their present jobs. Brent believed it was impossible to estimate what should be a future safe level of exposure for them.[24] A decade later, deVilliers and other experts testified that it would have been pointless to have transferred underground employees to surface jobs, since they had already been exposed to enough radiation to put them at risk for developing lung cancer.[25] The low doses of radiation they were exposed to underground after 1960, he said, were too small to make any difference. At that time, the government was only being prudent in expressing no opinion. Had it insisted on transferring all underground employees to surface jobs, the mines would have had to close down, and recruiting a new lot of miners, given the publicity of what had happened, would have had little likelihood of success. This would have meant the end of fluorspar mining at St Lawrence, with ruinous consequences for the community – a prospect no government would want to face. But neither could the government counsel the underground workers to continue at their jobs for fear of being held responsible should they subsequently develop lung cancer. Each underground worker had to decide for himself what to do, some after consulting their physician.[26] In fact, the dilemma was largely resolved by the workers themselves. In the years to come, there was a steady attrition from the ranks of the pre-1960 underground workforce; by 1967, there were only seventeen of the original seventy who had started in January 1959. That number further shrank to thirteen by 1975.[27]

On account of the widespread rumours among the workers and in the media that the mines were still not safe, Newfluor began to experience a high turnover rate of employees and difficulty attracting replacements.[28] Manpower shortages become so acute that, in 1967, Newfluor's fluorspar production fell short by 30,000 tons of the 150,000 tons required by Arvida.[29] However, in the next year the number of applicants showed a gratifying rise.[30]

The sixties saw a further expansion of Newfluor's holdings.[31] New shafts were sunk in Director. The St Lawrence Corporation's assets were purchased in 1964 as a reserve but also to reopen Tarefare. There were several wage increases but never at par with the other two mining companies on the island.[32] For example, in 1968 wages were 16 per cent lower at St Lawrence than at ASARCO.[33] There were fringe benefits, such as a bonus for underground employees and three weeks vacation with pay for those with more than ten years of service.[34]

PERMISSIBLE EXPOSURE

The future of the St Lawrence fluorspar mines depended on keeping radon and radon progeny at safe levels. In 1960, there were no statutory permissible exposure levels for alpha radiation in uranium mines, neither in Canada nor in the United States, apart from the state of New Mexico.[35] In the United States, there was a number of proposed safe exposure standards in circulation but none had received the stamp of approval of Congress. That was to come in

1967.[36] When it came to setting standards, all the experts had to go on was the data accumulated on the Erzgebirge mixed metals and pitchblende (uranium) mines and the early reports on the Colorado Plateau uranium mines.[37] In 1941, in the United States, using data from the Erzgebirge Mountains mines, a workplace safety standard for radon was developed. Citing differences in uranium mines between the United States and Europe, the American authorities chose not to adopt this standard in their uranium mines. Instead, they decided to develop their own safety standards using data collected from their own uranium mines.[38] Uranium mining on an extensive scale began in the United States only after World War II, in the south western states.[39] In 1949, the US Public Health Service (PHS) initiated long-term environmental and epidemiological studies in the Colorado Plateau uranium mines.[40] Because of the long latency of radiation-induced lung cancer, (estimated on an average to be about seventeen years), it would probably be at least several decades before a statistically significant increased rate of cancer of the lung could be detected.[41] In fact, it was in 1960 that the PHS informed the governors of the uranium mines states that, indeed, they were facing an epidemic of lung cancer in their uranium mines.[42] However, earlier, in 1951, high radon levels, comparable to those observed in the Erzgebirge Mountains uranium mines, had been recorded.[43] This certainly had sounded alarm bells, but nothing had been done because of inability to agree on whose responsibility it was to take action. In the same year, it was shown that efficient ventilation was extremely effective in dispersing alpha radiation, and a group of public health officials had recommended that uranium mine operators aim at augmenting ventilation so as to reduce radon to a working level (WL) of at least 10^{-9} C/L.[44] Some operators voluntarily installed extra ventilation. Such was the status of alpha radiation safety in the US uranium mines when the Newfoundland government was confronted with the urgent task of setting its own permissible exposure level to alpha radiation for the first time in mines in Canada.[45]

The Newfoundland government looked to National Health and Welfare, specifically Windish, for guidance. Since the federal and provincial governments had yet to determine what was to be the role of each in regulation,[46] he recommended what was widely followed in the United States for alpha radiation in uranium mines – the Holaday Interim Working Level (WL) of 1957. It stated that a worker should not be exposed anytime to more than 1 WL.[47] This recommendation assumed that carcinogenic risk from exposure to alpha radiation was threshold, safe below a certain level. About this time, this assumption was beginning to be questioned certainly for the delayed effects from exposure to alpha radiation, such as carcinogenesis.[48] Unlike the prompt effects where severity is dose dependent, cancer is an all-or-nothing phenomenon – you develop it or you do not. Thus, risk became probability estimates for developing cancer in response to different amounts and rates of exposure to alpha radiation. In some respects, a linear model fitted the facts better than a threshold one.[49] One determinant of risk is cumulative dose. Another is rate of accumulation.[50] The Holaday Interim 1 WL took no account of such complexities in standard setting for permissive exposures to alpha radiation.

To do so, another type of measurement, a personal one for individual cumulative exposure to alpha radiation and its rate was devised – the Working Level Month (WLM). Soon WLMs superseded WLs in designating statutory permissible exposure limits to alpha radiation. When it came to individual protection, eventually Newfoundland followed the US Federal Radiation Council recommendation, approved by the president 27 July 1967, prescribing exposure to an upper limit of a total of 6 WLM in any consecutive three-month period and no more than a total of 12 WLM in any consecutive twelve-month period.[51] The important point to note is that under this regulation short-term excursions above 1 WL, not high enough to produce prompt adverse effects, were no cause for alarm – a point that both the minister of health and union leaders were unwilling to accept, or more plausibly unable to grasp.[52]

During the 1960s, a two-pronged scheme to provide safety from alpha radiation was followed. The first tracked the accumulating WLMs for each miner to ensure that they did not exceed the prescribed total WLMs for the statutory three- or twelve-month period.[53] The second, though not stipulated, closed work areas with locational WLs of one or higher.[54] In fact, the policy was to aim at keeping the WLs as low as reasonably achievable (ALARA). During these years, they were generally less than 0.5 WL.[55] Furthermore, no employee ever seems to have been removed from work because his WLM reached the stipulated amount for that particular three- or twelve-month period.[56]

FREQUENCY OF MONITORING

If setting safety standards had to be left to the experts, the same could not be said of frequency of monitoring measurements; here the miners and their advocates had definite opinions, ones they had no hesitation in expressing forcibly. The union demanded a frequency of monitoring that management thought unnecessary.[57] Consequently, establishing how often monitoring readings should be rationally taken was to prove a contentious issue. In 1959, Windish had suggested that once the additional ventilation systems had been installed and a general pattern of normal readings established, surveillance readings should be no less than every three months.[58] He offered no rationale for this recommendation. As of 1 April 1961, the chief inspector of mines ordered a sliding scale for frequency of readings, depending on the previous WL, at the then thirty-six working locations.[59]

Following the discovery of high alpha radiation levels in the mines, Newfluor's safety officer was directed to take weekly alpha radiation measurements at all workplaces in both mines. It took him all day to perform twenty readings at Director and half a day at Tarefare. If a reading measured 1 WL or higher, he testified later before the Royal Commission, the miners would be evacuated and the cause investigated and corrected. They would not be allowed back until the reading was 1 WL or less.[60] As a result of an agreement between the union and management, two employees were trained to use the company's instruments for measuring dust levels and radiation. Every two weeks one of them would accompany the safety officer when he was taking his dust and radiation measurements to verify his readings.[61] Every three or four months a government radiation monitoring technician would take dust and radiation readings with his own instruments and compare them with the company's.[62]

Although these radiation safety surveillance measures were in line with those elsewhere, they did not satisfy the union. It called for daily readings at every workplace by a government radiation monitoring technician because it doubted if the two workers trained in the use of Newfluor's machine would ever become fully proficient in its use.[63]

SAFETY CONCERNS

Successive union presidents complained vociferously to the minister of mines that in spite of the additional ventilation equipment, the miners were still experiencing exposures to dangerously high levels of radiation, and they bombarded the minister with their complaints on this score.[64] The exasperated minister reached the point of refusing to have any further dealings with one president unless, as he said in a letter to the union's solicitor, "his complaints are specific and verifiable [and] his solutions reasonable and applicable." The minister went on to add that in the opinion of the inspectors the mines were safe.[65] The discrepancy between the union leadership's public announcements on radiation safety in the mines and rebuttals from the experts and management calls for an explanation. It is accepted that the words and actions of each of the parties drawn into dealing with an industrial risk are dictated by their own specific interests. Viewed in this light, the sometimes extreme language of the union leadership at St Lawrence becomes understandable. Setting safety standards was bound to be contentious in the circumstances prevailing at St Lawrence in the 1960s. Establishing safe exposure limits to radiation was still at an early stage. Windish's recommendations were interim. Those who were not experts had trouble understanding that it was the Working Level Months (WLM) to watch, not Working Levels (WL), which is a stationary measurement.

Then there was the ideological climate governing relations between management and labour expressed in terms of class conflict. Labour was expected to adopt an adversarial pose towards management. The fate of the previous trade union at St Lawrence was a warning to the executive not to appear too accommodating to management. Behind closed doors it was different. There, the two parties had to hammer out agreements; but in public the union's language had to be highly critical of management. The charge that the union was engaging in a certain amount of posturing is therefore not unjustified. This was brought out some years later in an exchange between S.T. Payne, vice-president of the Confederation of National Trade Unions (CNTU), with which the St Lawrence Workers Protective Union was affiliated, and Judge Fintan J. Aylward, chairman of the Royal Commission on Radiation. Under relentless questioning, during which Payne persisted in reiterating his charge that no one could know for sure that the mines were safe, Aylward finally pinned him down and forced him to admit that, in the light of existing knowledge, the mines were safe.[66]

The union's leaders' charges did not remain an in-house matter; they spilled over into the media where they were taken at face value, and onto the floor of the Legislature, where naturally they were seized upon by the opposition as an opportunity to belabour the government. Smallwood's assurances that the mines were now safe were unable to stem the growing public unease about this question.[67]

PROBLEMS IN DIAGNOSIS

Questions about safety in the mines were not the only criticisms the authorities had to face. Adding cancer of the lung due to radiation to the Workmen's Compensation Schedule of Industrial Diseases entitled to compensation was so obviously a just measure that it raised no dissent. However, before Workmen's Compensation could accept a claim for compensation it required proof of diagnosis, which involved demonstrating the presence of cancer cells. During life this required either an examination of the sputum or a more reliable lung biopsy. Worker's Compensation sometimes encountered resistance when a claimant was asked to submit to a lung biopsy because workers often witnessed comrades die shortly after the removal of a lobe or entire lung and assumed all chest surgical interventions carried the same risk.[68]

Another consequence of adding cancer of the lung to the Workmen's Compensation Schedule of Industrial Diseases was the encouragement it gave to the belief that any chronic condition a miner might develop had to be work-related and therefore should be compensable.[69] When claims for such conditions were rejected because they were diagnoses that were not listed in the Workmen's Compensation Schedule of Industrial Diseases due to insufficient epidemiologic evidence, the decision was greeted with disbelief and anger.[70]

ECONOMIC CONCERNS

The Workmen's Compensation Board also had to cope with a complaint inherent to pension schemes: survivor pensions were less than total disability pensions of living wage earners. In St Lawrence, the drop in the family's income when a workman who was on full compensation died was a cause of much resentment.[71] However, a drop in income did not always occur. By the days of the Royal Commission in 1969, in addition to a lump sum and burial expenses, Workmen's Compensation Board paid the widow a monthly sum and an allowance for each dependent child, up to a total of seven children. This was the maximum possible total survivor pension for a widow with children at that time.[72] However, she had the option of foregoing applying to the Workmen's Compensation Board and applying to the Department of Welfare for total financial support. Welfare had no ceiling on allowances for dependent children.[73] Thus, it was to the advantage of a widow with up to seven children to apply for the more generous incomes of Workman's Compensation. With eight children, payments were the same from Workmen's Compensation Board and from Welfare. With nine or more children, the incomes from Welfare became progressively more generous with each successive child. Thus, it was to the financial advantage of widows with nine or more dependent children to follow the Welfare route. By so doing, depending on the number of children they had, they could end up with larger incomes than their deceased husbands' total disability pensions.

FUNDING WORKMEN'S COMPENSATION BOARD

When cancer of the lung due to radiation was added to the Schedule of Industrial Diseases entitled to compensation, Workmen's Compensation Board was faced with the necessity of having sufficient funds at hand to compensate what was anticipated would be a sizable number of claims. Workmen's Compensation Board derived its funds from the annual premiums each industry was required to pay. For assessment purposes, industries were grouped into sectors, each sector representing a particular industrial activity. Every industry in a sector paid at the same rate per payroll employee. The assessment for each sector was set by estimating the anticipated amount of disbursements on claims from that particular sector in the coming year plus an extra charge to go into a reserve disaster fund to pay for unforeseen calamities. Newfluor, along with the other mines and the one limestone quarry in the province, paid into the mining sector. When faced with what could be a considerable run on its funds to pay for the St Lawrence tragedy, Workmen's Compensation chose not to increase the assessment of the mining sector but to pay all claims for cancer of the lung due to radiation out of the disaster fund.[74] The reason was so as not to impose a heavy burden on the mining companies of the province but to spread the cost among all the industries. Another reason was not to frighten mining companies away from the province. The disaster fund did not become depleted in the 1960s, and, in 1967, it stood at almost the same level as in 1960.[75] By the mid-1970s, however, the fund was exhausted.[76]

GOVERNMENT RESPONSES

By the late 1960s, with mounting public concern about the safety of the mines,[77] the government was forced to take some action to show that it took these concerns seriously. One measure it took, in May 1967, was to place a full-time government-salaried radiation monitoring technician in the mines to take daily readings at all workplaces and to compare his readings with the company's readings, thus to show, as the government claimed, that the company's were accurate.[78]

The other decisive step the government took had its beginnings in a result of a regulation governing the operation of the Workmen's Compensation Board. Back in 1964, as was required every five years, the government appointed a committee to review the operations of the board and to make recommendations.[79] The committee, named the Winter Committee after its chairman, Judge H.A. Winter, among its other activities, examined the handling of claims from the St Lawrence mines. Rennie Slaney, a retired Corporation foreman presented a brief to the committee. His submission was later tabled in the Legislature and published.[80] It is a detailed account of conditions in the fluorspar mines in the early days. He also compiled a list of names of deceased miners whose deaths he believed could be attributed to their work. When it came to the radiation hazard, Slaney was outspoken in his concerns about radiation safety in the mines and the reliability of the ventilation system. His brief made a deep impression on the committee and indeed later on the Legislature. When it came to the St Lawrence fluorspar mines the committee's report had this to say: "The only recommendation we feel we can make is that this Act [Workmen's Compensation Act] be ignored altogether ... The tragedy ... has far exceeded the scope of the Act. On the grounds of cost alone it is impossible to apply."[81] It went on to predict that the industries in the province collectively would be unable to bear the cost of compensating claimants. (By 1964, the numbers of total and partial disability pensions at St Lawrence were eighty-three workers and ninety-nine widows.[82]) It concluded by saying, "The only positive recommendation we can make ... is that the whole history of fluorspar ... from its start in 1933, be investigated and examined … in its every aspect. What form the inquiry should take, it is for the proper authorities to decide, whether … a Royal Commission (RC) or in some other way."

The government tabled the Winter report in the Legislature early in 1967.[83] The report unleashed a furor with calls for an inquiry from all sides. Smallwood stuck to his guns, arguing that the mines had been perfectly safe since extra ventilation systems had been installed. He saw no need for an inquiry. Then, in words strangely tactless from such an astute politician, he added that the tragedy of the lung cancer epidemic was "passé, out of date, outmoded, and past history."[84] These insensitive words only further angered those calling for a full-scale investigation.[85] Although Smallwood feared that constituting a Royal Commission to investigate the St Lawrence tragedy could land him in trouble, and he did as much as he could to get out of doing so, his hand was forced. His minister of justice, T. Alec Hickman, was also the MHA for Burin, the constituency in which St Lawrence was located. After reading the Winter report he realized there would have to be a Royal Commission of Inquiry; his constituents would expect no less.[86] Smallwood, resigned to the inevitable, had come to the same conclusion. The necessary steps would now have to be taken to nominate a members list, set terms of reference, and provide resources and personnel for a Royal Commission to fulfill its mandate. Before turning our attention to the Royal Commission, however, the story of the Corporation requires completion.

5

The St Lawrence Corporation: The Twilight Years

The prosperity the Corporation had enjoyed during the war years came to an abrupt end in 1945. The Corporation had become heavily dependent on US contracts, which ceased with the war's end. Soon the Corporation, with no sales in the offing, found itself facing a financial crisis. No help could be expected from the Commission of Government. Walter Seibert pinned his hopes on the new regime.

FINANCIAL WOES

Within a few months of assuming office, Premier Smallwood was approached by Seibert, who painted a bleak picture of the finances of his company.[1] The reasons for this depressing scenario were foreign competition, low world prices, and crippling expenses.[2] Once again, the high cost of electric power and its unreliability were stressed as needing to be rectified to bring costs down.[3] This would require adding to the number and power of supplementary diesel generators, expanding hydroelectric power capacity, and putting a freeze on hydroelectric rates.[4] Banks were demanding repayment of loans and refusing to issue further credit. Unless there was an infusion of a line of credit of $250,000, the Corporation would have to slow down and operate only when it had orders to fill.[5]

Faced with impending bankruptcy of the Corporation, the government called in two auditing firms, one American and the other Canadian, to provide assessments of Seibert's Delaware and St Lawrence companies.[6] The auditors stressed what everyone already knew: the Corporation's fluorspar was not competitive on world markets. Both companies were one-man shows with Seibert holding 81–82 per cent of the shares in the Corporation and 65 per cent of those of his Delaware company, the St Lawrence Fluorspar Inc.[7] The St Lawrence plant was on land leased by the government to Seibert, who in turn leased it to the Corporation at one dollar per ton of fluorspar sold.[8] Peat, Marwick, and Mitchell and Co., the Canadian firm that audited the Corporation, considered the amount of credit requested by Seibert realistic, but with one significant qualification: "There is a considerable element of speculation in the future prospects of this company which would render a loan unattractive ... there being insufficient risk capital invested in the undertaking."[9] The auditors however understood that political considerations, such as the need to provide work in St Lawrence, might outweigh prudential reservations about the long-term prospects of the Corporation. But, should the government guarantee the loan (by law it could not give a loan),[10] it should insist that the profits go towards reducing the bank loan.[11] On 2 February 1950, the government guaranteed a line of credit of $250,000 from the Royal Bank.[12] The loan gave the Corporation respite.

US CONTRACT

The outbreak of the Korean War in June 1950 brought a further reprieve. Fearful that this war might be the opening salvo of a wider conflict, the US government saw an urgent need to stockpile large amounts of fluorspar.[13] American producers could not supply sufficient amounts, so it turned to foreign suppliers, including the Corporation. The US Defense Materials Procurement Agency called in the managers of Seibert's two fluorspar companies in July 1952 and negotiated with them to provide 150,000 tons of fluorspar, to be delivered over the next four years, beginning in 1953.[14] The agreement stipulated that the fluorspar had to be shipped in a submetallurgical (70% CaF_2) form to the Wilmington, Delaware plant for further upgrading to acid grade.[15] Accordingly, the agency advanced a loan of $1.25 million for the construction of new facilities in St Lawrence and Wilmington to meet the production goals. This included the construction of a heavy media separation plant in St Lawrence. The downside of this agreement was that all the fluorspar produced by the Corporation had to go to Wilmington (presumably duty-free). It was the end of supplying eastern Canadian steel mills with metallurgical fluorspar. It was also putting all of the Corporation's eggs into one basket.

During World War II and in the immediate post-war years, the Corporation had been the major provider of metallurgical grade fluorspar to eastern Canadian steel mills but then had begun to lose ground to foreign competitors.[16] By 1950, of the total amount of metallurgical fluorspar used in Canada, 1,572 tons, (27 per cent) was imported. With the prospect of losing its Canadian markets, it is no wonder, as Seibert explained to Smallwood, that but for the US contract, the Corporation would have been obliged to suspend operations.[17]

The Corporation exceeded the amount of fluorspar it was required to ship to Wilmington under the agreement with the US government (see Appendix B) and made a profit of $10 million.[18] Seibert used the Korean War years to develop his fluorspar holdings in the United States, expanding the Wilmington plant and building a drying plant in Cleveland, Ohio. He also purchased a two-thirds interest in a Mexican fluorspar mine and its flotation mill; the product of this Mexican mine went almost entirely to his Wilmington plant.[19]

SUSPENSION OF OPERATIONS

With the expiry of his US contract and no customers in the offing, Seibert had little choice but to suspend his St Lawrence operation in June 1957.[20]

Looking around for markets, Seibert had foreseen that with a US duty of $8.40 per ton for metallurgical grade fluorspar and $2.00 per ton on acid grade, and lobbyists pushing for further increases in duties, the United States was not a hopeful market.[21] Canada, on the other hand, had no tariff on imported fluorspar.[22] If the federal government could be persuaded to stockpile or to levy a tariff on imported fluorspar, the Corporation might be able to recapture its Canadian markets.[23] As Seibert had put it earlier to Smallwood, "I am in the paradoxical situation of trying to prevent an increase on duties in the United States yet urging Canada to levy a tariff on imported fluorspar."[24]

CANADA TARIFF BOARD

In July 1957, Seibert wrote to G.H. Glass, deputy minister of finance in the federal government, requesting a tariff of at least ten dollars a ton on imported fluorspar, citing foreign competition as the reason.[25] He realized that steel, glass, and ceramic users of fluorspar would object, but considering the small amounts of fluorspar they used in their industrial processes, the additional costs, he reasoned, would not be a burden. Since Seibert now proposed selling all three grades of fluorspar, he would have to reactivate the froth flotation mill at St Lawrence to process them. The impending opening of the St Lawrence Seaway, he pointed out would reduce transportation costs to the Great Lakes industrial heartland of the continent. Failing a tariff, he concluded he would have to close.

In response to Seibert's letter, the federal minister of finance directed the Tariff Board to "make a study" of the fluorspar industry in Canada and the effects of the free entry of foreign fluorspar.[26] Industries using or mining fluorspar submitted presentations to the board. Surprisingly, Seibert and Poynter, vice-president of operations at the Corporation both submitted briefs to support the case for a tariff.[27] (Both extolled the Corporation and its treatment of its employees. In its later years, although the wage scale was less, workers were said to be less regimented and happier at St Lawrence than at Newfluor.[28])

Poynter's was the more detailed of the two submissions, providing a picture of affairs at the Corporation at that time from the point of view of management. The Corporation had always operated in the black, except for 1951. Labour relations were excellent with no work stoppages in eight years. Worker turnover was low. The vast majority of those who had been hired in the earlier days were still with the company. Wages for non-management employees ranged from $1.20 to $1.75 per hour for a forty-four-hour week. Wages accounted for 25 per cent of production costs. The company provided non-contributory life insurance policies ranging from $2,000 to $7,500, retirement policies on a fifty-fifty contribution basis, annual paid vacations of two weeks, and sick leave with pay. The Corporation had encouraged employees to build their homes with interest-free loans. Living in their own mortgage-free homes, close to their place of work, offset the disadvantage of lower wages compared to other miners, so Poynter argued.

Poynter then went on to enumerate the reasons for the high production costs. Naturally, at the top of the list was the high cost of running the pumps to keep Iron Springs and Blue Beach in a workable condition. If the Corporation was going to re-enter the Canadian market, it would have to be as suppliers of metallurgical grade fluorspar to the steel mills of eastern Canada. Refining fluorspar to metallurgical grade was wasteful because it generated "fines" that had to be removed before the metallurgical grade was fit for steel smelting. As there was little demand for acid grade fluorspar in Canada apart from Alcan, "fines" should be converted to acid grade for sale in the United States. This would require the reactivation of the froth flotation plant. Seibert estimated the Corporation could produce metallurgical grade fluorspar at twenty-five dollars per ton with added shipping costs of three dollars per ton to Sydney, Nova Scotia, or $5.50 per ton to Hamilton, Ontario. This was still higher than foreign competitors. Therefore, putting the Corporation in a competitive position would require a tariff of ten dollars per ton on both acid and metallurgical grades.

Hastings Company, which was at that time mining fluorspar at Madoc, Ontario, supported the Corporation's request for a tariff.[29] Elsewhere, the Corporation encountered opposition from all quarters to its plea for a duty on imported fluorspar. The steel industry was hostile, even though Seibert argued a tariff of ten dollars per ton would only add ten cents to the production of each ton of steel.[30] Alcan was also unsympathetic because so much of its aluminum was sold abroad, including to Mexico. It feared retaliatory tariffs as well as a rise in cost of its own production.[31] Seibert suspected that Alcan wanted to maintain the option of closing Newfluor one day and importing cheap foreign fluorspar, duty-free.[32] The Smallwood government was facing a dilemma. Although instituting a tariff lay within the jurisdiction of the federal government, the provincial government was, nevertheless, in the position of being able to apply some pressure on behalf of the Corporation's request, which Seibert urged Smallwood to do.[33] But to do so ran the risk of alienating Alcan. Still, if the Corporation was to be saved, something would have to be done quickly or skilled personnel would drift away.[34]

The provincial government was not so naïve as to accept Seibert's and Poynter's argument that a tariff would enable the Corporation to keep afloat. It needed to find out for itself whether the Corporation would be viable if protected by a tariff on imported fluorspar. Wisely, the provincial economist G.K. Goundrey was directed to prepare a report on the Corporation.[35] His report painted a gloomy picture of the Corporation's viability. Goundrey concluded, first, that to operate profitably the Corporation would have to produce 75,000 tons of fluorspar a year. The Canadian market, other than Alcan, was 30,000 tons a year. (He did not address the question of what was to be done with the excess fluorspar.) With the United States tariff as high as it was, it would be difficult to enter that market. Second, Goundrey concluded that the Corporation had done little in the way of exploration. Third, he noted, the mining methods were outdated and the equipment poor. For these reasons, Goundrey did not see how a tariff would alleviate the Corporation's present uncertain situation.

The union was concerned at the suspension of operations and rumours of closure. The president cabled Seibert asking if these rumours were true and adding, "have you forgotten your faithful employees?"[36]

In February 1958, pending the Tariff Board's ruling, the government called a conference of the interested parties and experts to discuss the Corporation's plight – the reasons for it and what could be done to save it.[37] Smallwood opened the meeting by pronouncing the suspension of operations by the Corporation as unjustified, but he provided no reason for saying so. Seibert repeated the reason – high cost of production – which compelled suspension of operations. He mentioned that there was an oversupply of fluorspar in world markets and prices had dropped. For the first time, Seibert referred to the high silica content of St Lawrence fluorspar as an unfavourable feature since the acceptable percentage of silica demanded by steel manufacturers had been steadily dropping since World War II.[38] Refining fluorspar down to these lower levels of silica was adding to production costs. He concluded by saying that, while a tariff might benefit the Corporation, what was really needed would be for the federal government to purchase fluorspar to the tune of $250,000 and stockpile it for sale when demand would pickup. This would provide work for fifty to seventy-five men. The attorney general claimed stockpiling would only make sense if it could eventually be disposed. The provincial economist remarked that the fundamental problem was cheaper Mexican fluorspar, which was already being shipped to Seibert's Wilmington plant. Gover, the deputy minister of mines, dwelt on the technical deficiencies – old equipment required repeated repairs and therefore caused slowdowns. The failure to conduct exploration meant no one knew the extent and grade of remaining deposits, and, consequently, there was no development plan. When a vein ran out, one looked in all directions for a new one worth mining.

The conference concluded on the sombre note that, in spite of Seibert's ingenuity and the workers' dedication, the Corporation had only survived this long because of abnormal circumstances. Its time had now run out. Nonetheless, the government persisted in trying to make the Corporation viable. It prodded Seibert to reopen Iron Springs by offering to guarantee a loan of $75,000.[39] What became of this offer, however, is not recorded.

CANADIAN TARIFF BOARD'S REPORT

In September 1958, the Tariff Board presented its report to the federal minister of finance.[40] It enumerated the unfavourable circumstances besetting the Corporation. It zeroed in on the impossibility of the Corporation lowering the silica content of its fluorspar except by wasteful and expensive refining. Even with a 10 per cent import duty, Mexican fluorspar would still be cheaper upon delivery compared to the Corporation's. The lower levels of silica in Mexican fluorspar made it more desirable for smelting steel. Labour costs were 35 to 50 per cent of total costs, but it was other factors that were contributing to the high cost of St Lawrence fluorspar, contrary to Seibert's claim that it was cheap labour that was giving foreign fluorspar the advantage over St Lawrence.[41] Alcan, which exported 80 to 85 per cent of its aluminum, could not afford the risk of this being jeopardized by a tariff. St Lawrence fluorspar was not suited for steel smelting. It would be more practical for the Corporation to concentrate on producing acid grade if markets could be found. But finding markets was unlikely because Newfluor supplied almost all the acid grade that Canada required. For these reasons the board could not recommend a tariff.

THE END OF THE CORPORATION

Seibert and Poynter accepted the board's decision with good grace. They seemed to have clung to the hope that the suspension of operations was temporary, and, indeed, some intermittent mining was carried out during the next few years. In November 1958, Seibert asked the government to guarantee a loan of $200,000 to go towards stockpiling.[42] Nothing seems to have come of this request. Surprisingly, in the summer of 1959, St Lawrence metallurgical fluorspar was being sold to steel mills on the Great Lakes at prices competitive to Mexican fluorspar.[43] The Corporation was able to do this, so Poynter said, by producing "two grades" of fluorspar from the refining process, one sold in the United States and the other in Canada.[44] Poynter said he had also been able to develop a small but steady market for "both grades" in the United States, employing a small number of men at St Lawrence. He was optimistic that eventually he could bring the Corporation up to full capacity. He had just returned from New York with his "pocket full of orders."[45]

However, the Corporation's days were numbered. On 6 June 1961, Seibert died.[46] His death removed the driving force that had kept the Corporation going. Ownership now passed to his heirs, who did not share the same enthusiasm for continuing the mines.

Another event must have raised questions about the wisdom of maintaining the Corporation – namely, the alarming number of cases of cancer of the lung that were beginning to be diagnosed in the underground workforces of both mining companies. The Corporation so far had managed to escape public attention because, since 1957, it had drastically scaled down its operations. This good fortune might not continue, especially since all the early cases of cancer of the lung, not surprisingly, had been employees of the Corporation.[47] Furthermore, Windish had found high radiation levels in the parts of Iron Springs mine that had not been flooded.[48] In spite of the mounting evidence that a serious health hazard was present in the mines, Poynter refused to be convinced. As late as May 1960, two months after the release of the government's press conference incriminating radiation as the cause of the high rate of lung cancer at the St Lawrence mines, he had this to say: "I still believe that the trouble was based on an abnormal rate of tuberculosis" (quoting the fifth item in the 1 March 1960 Governmental Statement).[49] Poynter then went on to add: "Last fall the Fed boys found radioactivity in some odd spots that exceeded levels considered safe. These results were not consistent and at times erratic, however did point to the need of further research ... In the one mine we operate, there was only one unused spot with a high level ... and no one had been there for years. It was of no importance."[50] Still, he added, the Corporation planned to install extra ventilation. Poynter's comments should be no cause for surprise. When an operator gets the first inklings that there may be a serious health condition in the workplace, because of possible legal liability and financial costs, the operator is going to go to great lengths to avoid acknowledging an occupational cause.[51] Conversely, workers are motivated to claim illness is work-related. What is surprising is Poynter's persistence in minimizing the problem in the teeth of overwhelming evidence to the contrary. He was not the only one. Leondard Miller, deputy minster of health, as late as 1967, was quoted as viewing tuberculosis as the major health threat in the St Lawrence mines.[52] In sharp contrast, Alcan, from the very beginning, appears not to have indulged in such wishful thinking.

Still, regardless of playing down the seriousness of the health problem in the mines, prudence did suggest to the Corporation to stay out of the limelight. As Poynter observed, "you will note we are keeping out of the publicity side of things and do not care to enter into any of the controversy. Most of our mines had such minute amounts of radioactivity that it has been decided that even if the instruments were not sufficiently accurate to measure it ... we certainly don't want to enter into all the problems that our other friends have had."[53] The policy of silence worked. During its remaining years the Corporation managed to avoid public attention.

Following Seibert's death, his family put the Corporation up for sale. Interested in developing Tarefare and reopening Blue Beach, which had collapsed in March 1957 due to rock falls, Alcan purchased the Corporation in June 1965.[54]

Becoming totally dependent on the US government to subsidize its operations and for access to American markets was the downfall of the Corporation. Still, the policy benefited both parties during wartime but was of no advantage to the United States in peacetime, when domestic sources were sufficient to supply industrial needs. In fact, though, Seibert had no choice but to accept this arrangement, all it did was postpone the inevitable end of the Corporation.

The St Lawrence Corporation also owed its very existence to the vision and drive of Seibert, an entrepreneur if ever there was one. He must have been a man of great persuasive gifts to have been able to talk his men into working under the most primitive conditions in the 1930s and to persuade them and the local merchants to accept such an erratic system of payment of wages and for goods. Fortune smiled on Seibert. How long his company would have lasted but for two wars can never be known, but these two events allowed him to bring up his operations to a reasonable level of efficiency and acceptable working standards.

He could not have done it alone. For many years his man on the spot was Poynter, an American engineer. It was during his years that conditions improved so vastly. Nevertheless, the workers thought Poynter unfeeling and grudging when it came to making concessions.[55] To others who were qualified to know how a mine should be run, he appeared to be a competent and capable operator.[56] No doubt, like anyone having to run a tight enterprise on a limited budget, he would not be popular with those having to work under him.

Newfluor was now the only party mining fluorspar at St Lawrence and the one that became the focus of the Royal Commission's attention.

6

The Royal Commission and the Aftermath

When a calamity such as the epidemic of lung cancer at St Lawrence occurs, the public will want to know who or what is responsible for the misfortune, what needs to be done to put things right, and how to avoid a recurrence. A Royal Commission of Inquiry is the appointed way for countries with British institutions to provide the answers to these sorts of questions. The government is free to accept, reject, ignore, and evade the recommendations of a Royal Commission.

The composition of a Royal Commission is crucial. The members are government appointees. In calamities where liability issues are paramount, the chair is usually a lawyer. The other members are stakeholders, one way or another, in the calamity. Royal Commissions are costly because the commissioners are dependent on paid experts for technical and scientific information. These features of a Royal Commission also characterized the Royal Commission that investigated the radiation at St Lawrence.

Premier Smallwood told his minister of justice, Alec Hickman, to recommend the chairman for the Royal Commission.[1] Hickman chose Fintan J. Aylward,[2] a lawyer and former president of the Liberal Party, the one that had governed under Smallwood since 1949. Aylward was a native of St Lawrence, and his brother, who owned one of the stores, was the mayor. On 5 April 1967, a Royal Commission to inquire into all aspects of the radiation hazard and compensation at St Lawrence was constituted.[3] The other members were H. Bliss Murphy, a radiologist and oncologist; W. David Parsons, an internist who had been a member of deVilliers's epidemiologic team; and Frederick Gover, a mining engineer and deputy minister of mines.

Questions were raised about the membership of this Royal Commission. The union would have liked to have been represented.[4] Alcan wondered if the chairman's past political affiliation with the governing Liberal Party might influence "the direction" of the investigation.[5] In retrospect, the appropriateness of appointing Gover to the commission can be questioned. He was deputy minister of mines (the department that shared with Health responsibility for OHS).

The program the Royal Commission set for itself was a busy one. Specialists in epidemiology, industrial hygiene (dust conditions), and radiation were interviewed. The commissioners visited St Lawrence to confer with the mayor and city councillors, the cottage hospital physician Brian Hollywood, and the president of the St Lawrence Workers Protective Union. Newfluor, the Town of St Lawrence, and the Workers Union presented briefs. Opportunities for cross-examinations were provided at open sessions in St Lawrence. Public notices inviting anyone who wished to appear before the commission at specified times were put up in St Lawrence and neighbouring communities as well as in the St John's newspapers. Rennie Slaney was retained to present a brief describing work conditions. All known claimants for compensation were contacted. The commission sought claimants' resumés or a prepared history of their exposure history and health. On the basis of claimants' resumés, the commission decided which were valid diagnoses of occupational lung cancer. The commission presented its report to the government in July 1969. In its report, the Royal Commission listed sixty-nine conclusions (C) or recommendations (R) in Section 1, Ch. 19 and a few more relating to safety, in Section 2, Ch. 4.[6]

RADIATION SAFETY

One of the commission's terms of reference was to report on alpha radiation safety levels in the mines. It noted there were no statutory prescribed exposure limits for underground miners in Canada, except in Newfoundland. Radiation levels in the mines from January to June 1968 averaged 0.325 WL or 3.90 WLM, lower by far than recommended permissible levels anywhere else in North America.[7] The Royal Commission endorsed without comment (C10) the government proposed permissible levels as of 1 January 1969, of 1.8 WLM in any consecutive three-month period or 3.6 WLM in any consecutive twelve-month period.[8]

The Royal Commission cross-examined David Rex, the government radiation technician, stationed at Newfluor St Lawrence, about his monitoring practices.[9] He was supposed to perform daily measurements at every working area in Newfluor's two mines, but actually he could only do about twenty a day. Ninety-five per cent of his readings were less than 1 WL and were comparable to the company's readings. When asked how frequent readings should be, he replied daily because it would make the workers feel safer. This would require hiring another technician. Consequently, the Royal Commission recommended that two full-time radiation monitoring technicians be employed at the mines because the workers "are morally entitled to have every working place monitored daily and the results posted daily" (R14).[10]

CLAIMS TO WORKMEN'S COMPENSATION BOARD

Perhaps the most demanding charge handed to the commission was to investigate the processing of claims from the St Lawrence mines to the Workmen's Compensation Board (Appendix E) and whether its awards were sufficient. In its report, the Royal Commission found that the Workmen's Compensation Board "had been reasonably efficient in its handling of the majority of claims" (C65).[11] However, the Royal Commission expressed reservations about the impartiality of the committee of three medical referees, instituted in 1964, to handle appeals from decisions of the Workmen's Compensation Board.[12] The commissioners argued since these three doctors (J.B. Roberts [chairman], W.D. Parsons, and H. Bliss Murphy) were government appointees, an element of bias might enter into their decisions.[13] Therefore, the Royal Commission recommended that, in the future, each appeal should be heard by an ad hoc committee consisting of three specialists in the disease under consideration, one nominated by the appellant, one by the Workmen's Compensation Board, and the third, who would act as chairman, nominated by these two (R50). The Royal Commission took note of the reluctance of some claimants to submit to a bronchoscopy or lung biopsy, which delayed proof of a suspected diagnosis of lung cancer until autopsy (C35).[14] The commission also stressed the need for public education so that dependents could see that it was to their financial benefit to consent to autopsies to establish a diagnosis of lung cancer, preparing the way for compensation (C44).[15] However, in the absence of autopsy in suspicious cases, the dependents should be given "the benefit of the doubt" and awarded compensation (R24).[16]

At the time of the Royal Commission, the financial provisions of Workmen's Compensation for dependents were widely criticized as insufficient. They were those stipulated in the amendment of 1967 to the Workmen's Commission Act of 1962.[17] Workmen's Compensation benefits had been steadily augmented periodically since 1951 and were comparable to those of other provincial Workmen's Compensation Boards (see Appendix F).[18] The contentious point was whether survivor pensions were sufficient for the needs of widows and children. An independent survey reported that under "ideal conditions" they were "an adequate subsistence allowance" but insufficient to provide for extras such as secondary education for children.[19] Some of the widows supplemented their pensions by part- or full-time jobs, mostly in the service sector.[20] One, who became a bank teller, was able to provide university education for two sons.[21] Still, the Royal Commission could not take commendable initiatives such as these into consideration in its assessment of the economic status of the widows and their children. They were, the commission stated, suffering hardship and should receive "further financial assistance."[22]

Before it addressed how this should be done, the Royal Commission uncovered what can only be described as a singular state of affairs. As mentioned in chapter four, a widow had the option of going to either the Workmen's Compensation Board or to the Department of Welfare for financial assistance. With eight or less dependent children, it was to her financial advantage to apply to the Workmen's Compensation Board for support.[23] With more children, the benefits from social services were higher than from the Workmen's Compensation Board. To understand how this strange practice developed requires an understanding of the origin, purpose, and evolution of workmen's compensation.

Workmen's compensation first came on the scene in Germany in 1883 to deal with workplace injuries and accidents, and in the years that followed, it spread to other industrialized countries. Apart from poor relief, it was the earliest state-run relief scheme in the Western world. In the minds of its originators, it was to be a form of "no fault" insurance providing stipulated amounts of compensation for workplace accidents without having to resort to litigation.[24] Later, occupational illnesses were added to the list of compensable conditions. This was a step that has been a source of misunderstandings and bitterness between workers' compensation boards and their clients because of different conceptions of causation of diseases. The WCBs require scientific evidence before adding a disease to their lists of compensable conditions. On the other hand, there is a widespread opinion among the clients that any serious disease developing in a worker must have an occupational component.

However, according to its proponents, the Workmen's Compensation Board was never intended to be a form of social assistance. This is made clear by Justice W.D. Roach's comments in 1951, quoted as late as 1972 by the Newfoundland Workmen's Compensation Board with approval: Workmen's Compensation is not to be a scheme for dispensing charity nor raising the standard of living. If its disbursements are insufficient to provide a decent standard of living, then it is up to society to make up the difference through social assistance.[25] He could have added that, unlike the Department of Welfare, which handled its clients on a case-by-case basis through social workers, Workmen's Compensation was not structured to do so. At St Lawrence the Workmen's Compensation Board and the Department of Welfare, according to Roach's way of thinking, should have been working together as an integrated team on each claimant's case; they were not. Instead, claimants were going to either Workmen's Compensation Board or Welfare without guidance as to which was to their financial advantage. By not going to Welfare, widows on Workmen's Compensation were also depriving themselves of discretionary benefits, such as lump sum of $2,400 toward housing and exemption from paying municipal taxes, which were granted on a case-by-case basis.[26] The Department of Welfare told the Royal Commission that widows on Workmen's Compensation could have applied to the department and would have been entitled to all its benefits less any sums they had received for the same item from other sources.[27] There seemed to be general lack of awareness among the widows on Workmen's Compensation about their rights to discretionary benefits from the Department of Welfare. The Royal Commission took note of this lack of coordination between the Workmen's Compensation Board and the Department of Welfare and recommended that widows on Workmen's Compensation be made aware of their rights to extra benefits from the Department (R52). It also recommended removal of the ceiling on the number of dependent children for which a widow received allowances from the Workmen's Compensation Board (R64).[28]

The Royal Commission chose not to advocate an integrated Workmen's Compensation Board and Department of Welfare approach for handling compensation claims from widows and dependent children of miners deceased from lung cancer. Instead, the Royal Commission shared the widely expressed opinion that, for the St Lawrence widows of victims of lung cancer, having to resort to welfare for financial support was "unjust, humiliating, and degrading."[29] This is one reason why preserving a distinction between insured services and social assistance breaks down and government-operated insurance schemes are pushed into fulfilling both roles.[30]

Therefore, the commission argued, the widows should be entitled to full support from Workmen's Compensation.[31] With the removal of the ceiling of eight children, there would no longer be an incentive for resorting to the Department of Welfare, except for its discretionary benefits. Even so, the commissioners believed additional financial support would still be required. In its submission to the Royal Commission, Alcan had recognized that this was necessary and proposed a special fund to bring up recipients of Workmen's Compensation or welfare benefits to an acceptable standard of living.[32] To qualify for support from the fund, the claimant had to be disabled from lung cancer, silicosis, or silicotuberculosis, or the dependents of such a claimant. The Royal Commission took this proposal but expanded it to include any disabled worker from St Lawrence on Workmen's Compensation or his dependents (R53). Furthermore, the fund should be financed in equal proportions by Alcan, the Seibert family and both the federal and provincial governments (R54).[33]

CONTRIBUTION OF SMOKING

The Royal Commission collected data on the epidemic of lung cancer. By the end of 1967, the number of deaths had reached fifty-two, of whom only two were non-smokers.[34] The average age at death was less than that for lung cancer in the general population.[35]

Speculation on the contribution of smoking to the epidemic of lung cancer could not be avoided. By this time, epidemiological findings on the Colorado Plateau uranium miners were beginning to be released, and the Royal Commission had the benefit of advice from several experts on the interpretation of these findings.[36] Not surprisingly, the incidence of lung cancer among the Colorado Plateau miners was higher than in the general male population. Of particular interest was the difference in cancer rates between smoking and non-smoking miners. For the latter group, the rate of lung cancer was less than in the never-smoked general population. Gene Saccomanno, the principal investigator reporting these statistics, concluded that there was a very potent synergistic relationship between smoking and radon and radon progeny, and that, if cigarette smoking were eliminated, "cancer of the lung would be nearly nonexistent in the Colorado Plateau."[37]

Was smoking as important a cofactor in the cancer of the lung at St Lawrence as it was on the Colorado Plateau? The overwhelming majority of the St Lawrence fluorspar miners had been smokers, mostly heavy.[38] Unfortunately, because of lack of data on comparison groups in Newfoundland, no statistical analyses could be performed to assess the contribution of smoking as a cofactor.[39] Nevertheless, the commissioners were convinced that "the evidence strongly suggests that there is a synergistic influence between smoking and radiation." They put their stamp of approval on cessation of smoking programs directed at current and former miners (C27).[40] Ever since the government's statement of 1 March 1960, Newfluor had initiated anti-smoking programs, and, by 1964, the company was congratulating itself on the number of miners who had quit smoking.[41] In the years to come, however, management became less convinced of the efficacy of these programs and disturbed by the nonchalant attitude of its workforce about the dangers of smoking.[42]

HISTOLOGY

No account of alpha radiation-induced lung cancer in mines would be complete without a discussion of its histology, the microscopic structure of tissues. Was there a distinct cell type? All of the three major cell types of lung cancer, epidermoid (squamous), small cell undifferentiated, and adenocarcinoma were increased among the American uranium miners.[43] The largest increase was in the small cell undifferentiated. The same cell type was also the predominant one in the Erzgebirge Mountains mixed metals and uranium mines.[44] In the lung cancer associated with cigarette smoking, the same three cell types were increased but it was the epidermoid that showed the greatest increase.[45] The report of the Royal Commission claimed that among the St Lawrence fluorspar miners who developed lung cancer, the majority were "oat cell," a subtype of the small cell undifferentiated.[46] It cited no figures to support this statement. Saccomanno postulated that while cigarette smoking is a potent cocarcinogen, it has little if any effect in determining the cell type of lung cancer.[47]

SILICOSIS

The Royal Commission interpreted its mandate to include an assessment of the other chronic respiratory diseases and the cancers afflicting the workforce, particularly the role of radiation in their development, and to recommend which ones should be compensable in addition to silicosis and cancer of the lung due to exposure to radiation. As noted in chapter three, the incidence of silicosis at St Lawrence, compared to other mining operations, was not high. By the time of the Royal Commission, Workmen's Compensation Board had accepted fifteen claims for silicosis.[48] Still, there were complaints that one year a miner would be told that his chest x-ray showed early silicosis and the next year that his chest x-ray was normal. There was a perfectly simple explanation for these contradictory reports. Silicosis is diagnosed by the chest x-ray appearance of multiple small round nodules, which slowly grow in size. At the very earliest stage of their appearance there are bound to be differences of opinion on whether they are present. To achieve greater consistency in x-ray interpretation, the Royal Commission endorsed the suggestion in the Newfluor submission to the Royal Commission that miners' chest x-rays be read by a single radiologist (C38).[49] The Royal Commission raised the question of a synergistic relationship between radiation and silicosis only to reject it on the grounds of the low incidence of silicosis in the St Lawrence miners.[50]

TUBERCULOSIS

Tuberculosis was also on the agenda of the Royal Commission. It was supplied with statistics, which showed (as mentioned in chapter three) an increased susceptibility to tuberculosis among the miners.[51] This increased susceptibility had to be related to mining experience. The Royal Commission report referred to the well-recognized synergistic relationship between silicosis and tuberculosis and suggested that this might operate at levels of silica inhalation too low to be detectable in chest x-rays. This would account for the increased incidence of tuberculosis without radiological evidence of coexisting silicosis.[52] As with silicosis, the relationship between tuberculosis and radiation had to be considered. Reference was made to the breakdown of quiescent tuberculous lesions following exposure to acute radiation.[53] However, the effect of chronic low dose alpha radiation on tuberculous lesions was unknown. Still, the commissioners recommended that, as in the UK, silicotuberculosis be compensable regardless of which developed first (R41). The commission also recommended that tuberculosis occurring in the underground fluorspar mines be compensable (R59, R63).[54]

CHRONIC OBSTRUCTIVE LUNG DISEASE

The last group of chronic chest diseases to engage the Royal Commission's attention is really a single entity beginning as chronic bronchitis progressing to chronic obstructive lung disease (COLD) and emphysema. At the second stage, progressive disability begins. Chronic bronchitis is common in the general population and is attributed to smoking, environmental pollution, or both. The prevalence of chronic bronchitis in the St Lawrence miners was above that of non-miners in the community but comparable to other mining populations not exposed to high radiation levels.[55] Yet, the Royal Commission, while admitting that there was no evidence to incriminate radiation as contributing to chronic bronchitis, was "satisfied that it might be a contributing factor."[56] Therefore, the Commissioners recommended that COLD in workers exposed to high doses of radiation should be compensable (R29 and R60).[57]

OTHER CANCERS

The commissioners could not escape expressing an opinion on whether high alpha radiation doses might not also cause cancers in other parts of the respiratory tract that have been exposed to alpha radiation, namely the nose, mouth, pharynx, and trachea. It admitted there was no evidence of such an association, but neither was it possible to prove there was not a connection. Therefore, cancers of the nose, mouth, pharynx, and trachea should be compensable (R62).[58] Apart from inhalation, the only other way radon and radon progeny could possibly induce cancer would be by ingestion. The Royal Commission cited reports of two communities with high cancer rates and high levels of radon in the drinking water.[59] In St Lawrence, high radon levels were found in the municipal water system and in several wells.[60] Nevertheless, mortality rates for cancers other than of the lung were similar to other communities in the province, except for an increase in stomach cancer.[61] As a precaution, then, the Royal Commission recommended replacement of the municipal water system by one with safe levels of radon (R33).[62] This was done.[63] Because the underlying granite bedrock contained a higher amount of uranium than was found in other granites, the commissioners had to determine if it posed a health risk.[64] A gamma radiation survey of the mine was commissioned, and the values were not found to be elevated.[65]

MINERS' HEALTH SERVICES

The Royal Commission next turned its scrutiny on the health care available to the miners in 1967. Like other rural Newfoundlanders, primary health care was provided for them and their families through the cottage hospital system, at St Lawrence through the local hospital. The commissioners were told that the care they received was held in the "highest regard" by the people.[66] This is hardly surprising since the system has been praised for the high quality of the care it delivered.[67] The two cottage hospital physicians at St Lawrence also functioned as company physicians, performing the pre-employment and annual miners' medical examinations. If they suspected a worker had developed an occupational disease, they referred him to Workmen's Compensation Board. There were occasional complaints by workers about the way they were treated by Workmen's Compensation Board physicians. Thus, the cottage hospitals' physicians' roles as family and Company physicians were entwined. At St Lawrence, the physicians must have been discharging both to the satisfaction of their patients and the companies because the Royal Commission had no criticism to make, apart from noting a few incidents of delay in communication.[68]

The report of the Royal Commission contains a history of the health services in St Lawrence. The story up to 1954, the year the St Lawrence cottage hospital opened, has been told in chapter two. Now it needs to be continued to the closure of Newfluor. C.J. Walsh, the physician who first noticed the increased incidence of cancer of the lung, left after a few years and was replaced by Brian Hollywood, an Irish physician, as senior medical officer. Unlike the junior staff physicians who stayed only for a few years, he spent his remaining working years at St Lawrence and indeed chose to retire there. In 1954, Alcan and the Newfoundland Department of Health negotiated an agreement whereby the cottage hospital physicians would attend to the occupational health requirements of Alcan employees in return for one dollar per worker per month, paid to the Department of Health.[69] In this arrangement, no one could accuse the physicians of being in the pocket of management – a charge that has been levelled at company physicians.[70] On the other hand, it was an arrangement that St Lawrence cottage hospital physicians understandably resented because none of the monies collected under this arrangement went into their pockets. Each was paid a salary depending on the post he held. Cottage hospital physicians were allowed to keep fees derived from non-contractual services such as those for Workmen's Compensation and the Department of Veteran's Affairs.[71] In defence of the arrangement at that time, the duties the cottage hospital physicians were called upon to perform for Newfluor were not onerous – pre-employment examinations and fitness to work certificates. However, in 1960 the workload of the cottage hospital physicians for Newfluor increased when annual medical examinations, at a fee of $7.50 per worker paid for by the company, became mandatory.[72] These fees continued to go into the coffers of the Department of Health but now they represented a considerable sum, adding to the physicians' resentments.[73] The physicians began to complain that their increased workload was too much to handle in addition to their other duties. Alcan was sympathetic to both their complaints.[74] Various solutions were tossed about, and one was for Newfluor to hire its own physician. To do so though would have required paying him a higher salary than Hollywood's, which the cottage hospital physicians might resent.[75] Presumably for this reason Miller was said not to be enthusiastic about the idea of a third, but company, physician in St Lawrence.[76] Eventually, the scheme adopted was that proposed by Wiseman.[77] Each year, Alcan would dispatch a team of nurses and a physician to perform the annual miners' medical examinations. This continued until 1973 when Alcan and Hollywood signed an agreement by which the latter undertook to perform pre-employment and the annual miners' medical examinations in return for an annual honorarium of $1,000 and $100 a week.[78] Alcan continued its payments to the Department of Health as stipulated in the 1954 agreement.[79]

The cottage hospital physician's dual roles of being a family physician and a company occupational physician for the same man could create dilemmas for the physician, which management appreciated and sought to get around. Alcan's solution was that, as an examiner for pre-employment and annual medical certificates, the cottage hospital physician would act only as an advisor to management, and management would be ultimately responsible for determining employability. As the plant manager explained, "He [the cottage hospital physician] would be in a difficult position if it were known he was the one who said you could or could not work."[80] Presumably, the examining physician discussed with management whether an employee was fit to work underground, and, if he was not fit, whether there were surface jobs he could perform. In this last case, a worker would be transferred to a surface job.[81] Hollywood sometimes seems to have found himself in the position of being obliged to issue a certificate of fitness to work underground in compliance with provincial regulations to a worker whom he thought was not fit.[82] Once again, Alcan was sympathetic to his quandary and discussed with him the feasibility of drawing up its own more strict medical standards for employability underground.[83]

Hollywood seems to have been able to perform his two functions as a family physician and a company doctor to the same people to the satisfaction of all parties concerned.

THE DR SUIKER INCIDENT

One incident threatened to mar this harmonious relationship. It arose out of a situation inherent to the cottage hospital system but was compounded by the mine's manager allowing himself to be drawn into the fray for personal reasons. In the late 1960s, Dr Alice P. Suiker opened an office as a private medical practitioner in St Lawrence.[84] This was a prescription for trouble. Conflict was almost inevitable when a fee-for-service physician sets up practice in a cottage hospital district staffed by salaried physicians. What hospital privileges was he (in this case, she) to be given? Who would look after her patients when she was not around?[85] To complicate matters, Suiker was married to M.E. Gooding, Newfluor's plant manager. Gooding could find himself drawn into any conflicts his wife might get into with Hollywood, which is what happened. Wiseman, the previous plant manager who had by this time transferred to Alcan's head office in Montreal, perhaps seeing a conflict of interest for Suiker's husband, had requested that Suiker stop doing miners' medical examinations, to which she agreed.[86] Soon conflict erupted between Suiker and Hollywood over her hospital privileges.[87] But it was a quarrel over the treatment of one of Suiker's patients that brought hostilities to a boiling point. Hollywood, apparently without notifying Suiker, discharged a patient she had admitted to the hospital a few hours previously. Gooding then drove the patient, a boy who had sustained an injury, to the Burin Hospital, where he was treated.[88] Gooding, later on his own initiative wrote an angry letter to Miller, the deputy minister of health, itemizing what he considered to be Hollywood's offences in the case and calling for his removal from St Lawrence.[89] In reply, Miller wrote that he "did not consider the withdrawal of Dr Hollywood as a satisfactory solution."[90] His reason for stalling is clear. As administrator of the cottage hospital system, retaining medical staff was a constant headache. When he found a dedicated, competent physician such as Hollywood, committed to staying in St Lawrence, and about whom there had been no complaints until Dr Suiker's arrival, he was not going to run the risk of losing his services. For Miller, this was the overriding consideration. Regardless of the rights or wrongs of the quarrel, Hollywood had to be kept in St Lawrence. Miller could not run the risk of instituting an inquiry into the dispute, one which might find in favour of Suiker. Therefore, to this extremely competent and astute civil servant, the best course was to do nothing and to avoid taking sides, trusting the problem would take care of itself. It did when Suiker and her husband a few years later left St Lawrence. Wiseman, when he had heard of Gooding's letter to Miller, thought it ill-considered.[91] With no support from head office, Gooding was powerless to press his case against Hollywood.

ROCK BURSTS

On 15 September 1967, some months after the Royal Commission had commenced hearings, a rock burst occurred in Director mine, killing three men.[92] As a result, the Royal Commission was given additional terms of reference to conduct an investigation into the cause of the accident and the safety of the mines.[93] Until this accident there had been four fatal accidents at Newfluor, none related to rock bursts or machines.[94] The Royal Commission concluded that this rock burst accident was due to a number of causes – inadequate supervision, a nonchalant attitude about taking risks, and failure to heed warnings.[95] The rapid turnover of underground workers since the discovery of high radiation levels meant less experienced miners and less sensitivity to incipient dangers.[96]

To assist it with the second of the additional terms of reference, the Royal Commission hired a mining engineer, G.R. Yourt, to report on the safety of the mines.[97] The accident prevention programs, he thought, were up to acceptable standards.[98] A safety committee with representation from management (one being the safety officer) and the trade union toured the mines regularly. Workers had the right to refuse to work in locations they believed to be dangerous and withdraw from a site when they sensed danger.[99] Yourt also scrutinized the accident frequency rates between 1957 and 1967 and declared that they compared favourably with other mines in the province.[100] The current mining methods, so the commissioners said, might be increasing the risk of rock bursts.[101]

DUST MONITORING

The Royal Commission looked at the dust monitoring measures in place at the mines from 1960 to 1967.[102] Dust levels were well below twenty million particles per cubic foot (mppcf), the safety level recommended by Windish and Sandeman in 1958.[103] Still, the commissioners expressed their opinion that dust sampling was not frequent enough and emphasized the limitations of the midget impinger, the instrument then used for dust measurements.[104] Newfluor continued to use the midget impinger because to have gone over to gravimetric sampling would have meant sending the samples to Arvida for interpretation. Management realized, though, that eventually it would have to take this step and also adopt personal dust sampling, the most reliable measurement of an individual exposure to dust.[105] When it came to measurements of personal exposure to dust, there was the hurdle of worker resistance to clear. Workers hated wearing personal samplers, claiming they were a safety hazard.

The other comments and recommendations of the Royal Commission on safety were of a "housekeeping" nature. Smallwood had been reluctant to the idea of a Royal Commission, foreseeing it could spell trouble for the government. Every now and then he would ask

Hickman, the minister of justice, how it was coming.[106] Assurances that it was progressing well earned the retort, he hoped it would never report.[107]

ROYAL COMMISSION CONCLUSIONS AND RECOMMENDATIONS

The chairman of the Royal Commission presented the final report to Premier Smallwood in July 1969. As mentioned previously, it contained conclusions (C) and recommendations (R).[108] The report is an invaluable source of information and statistics.

No one or no party was blamed for the tragic epidemic of lung cancer, but two government departments were criticized for their initial responses when asked to investigate what appeared to be a serious health hazard. The first criticism faulted the provincial Department of Health for "being unduly slow in its acknowledgement of serious occupational health hazards in the mines at St Lawrence" (C1).[109] The second censure, directed at the "federal Department of National Health and Welfare, criticized it for being unwarrantedly bureaucratic in that it would hear no provincial voice other than that of the provincial Department of Health in matters relating to the health of workmen in the mines at St Lawrence" (C2).[110] Regardless of whether these two censures were appropriate or deserved, they amounted to little more than slaps on the wrists. Leonard Miller, the deputy minister of health at the time the cancer epidemic began, retired in 1971 with his reputation unsullied. In fact, he was widely acclaimed and eulogized by Smallwood.[111] A very different story emerged a few years later in the tobacco and asbestos industries, where some stakeholders were accused of ignoring or suppressing early evidence of harmful occupational health effects.[112]

There were no class action suits, unlike in the United States, where a claim was filed in 1979 on behalf of a group of lung cancer cases, victims of conditions in the uranium mines in the 1940s and 1950s.[113] The suit was dismissed on the technical grounds that the US government is immune from prosecution. However, the court was emphatic that the miners' plight called for "redress" and Congress promptly passed a bill providing for compensation.[114]

It was not until the spring of 1970 that the provincial government began to announce its opinions and decisions concerning the Report of the Royal Commission.[115] The government was particularly concerned with monitoring radiation levels in the mines and compensation for industrial diseases and accidents.

A controversial safety recommendation required that, until personal monitoring was available, two full-time radiation monitoring technicians should be stationed in the mines; two was the number estimated to ensure that radiation measurements were performed daily at every working site (R14).[116] The reason given for demanding daily workplace readings was that the workers were "morally entitled" to this service.[117] By entitlement, presumably the commissioners meant a right to freedom from anxiety about safety, the same argument advanced by David Rex, the government's monitoring technician at St Lawrence who had also favoured daily readings.[118] The recommendation said nothing about whether the two technicians would be government or company employees.

By placing the reasons for daily workplace readings on moral grounds, the Royal Commission stifled public debate on the frequency of readings. No one would dare to stand up and to question the necessity for such a frequency of measurements. The Department of Mines had its own priority, which was to get out of the business of radiation surveillance and to saddle Newfluor with responsibility for implementing R14 and hiring two technicians (R14). Therefore, the government saw no need to take any action on R14, apart from expressing its opinion that radiation monitoring was the operators' and not the government's responsibility.[119] It expressed no opinion on how frequent worksite radiation measurements should be. Assuming that Newfluor would have to implement R14, Miller warned Wiseman that the company would be expected to monitor every workplace daily.[120] Wiseman noted this would not only require two full-time technicians but the union would demand that one of its members be taken on as a full-time radiation monitor to check on company readings. This Newfluor would not accept.[121] Newfluor seems to have seen no need to switch from weekly to daily workplace radiation measurements. Consequently, the government found itself criticised for not immediately implementing R14 by appointing a second radiation technician. It seems that it was believed that both technicians should be on the government's payroll,[122] but the government would not accept this arrangement. The Department of Mines, which was responsible for mines inspection, did not want to pay for a full-time radiation technician.

David Rex had been appointed as a temporary measure in 1967 because of the union's lack of confidence in the radiation measurements of the company technicians, who performed measurements on a part-time basis.[123] Rex's measurements had correlated with those taken by the company technician and showed that "safe levels had been achieved." Rex's mission had been accomplished, and now the company would be approached and asked to take him into its employment.[124] It was the position of the government that radiation monitoring, like other health and safety procedures was the responsibility of the operating company not the government's.[125] The government's role was merely to monitor compliance with legally-enacted health and safety regulations. The department was only following "generally accepted practices," it argued, regarding Rex's appointment as a special but temporary measure and not a precedent.[126] The government perceived that, if it assumed responsibility for radiation safety at Newfluor, it would be opening the door to other companies to ask government to provide health and safety monitoring services for their operations. The expense could be high. Others did not share this philosophy, especially the opposition. In the Legislature, there were calls for the government to appoint a second radiation technician immediately.[127]

The government vacillated in its commitment to the Department of Mines' principle that monitoring workers' health and safety was a company, not a government, responsibility, but at one point Smallwood indicated his intention of appointing a second technician.[128] Behind the scenes, he was reluctant to go along with the Department of Mines' principle because to do so could be construed as being anti-labour. (The St Lawrence Workers Protective Union had advocated government-funded radiation monitors.[129]) However, Smallwood told the minister of mines, W. Callahan, that if he could persuade the union and the municipal council to accept the Department of Mines' request for Newfluor to take Rex into its employment, Smallwood would see that Newfluor did so. With the assistance of Lord Taylor,[130] president of Memorial University, Callahan was able to convince the union and the municipal council to accept the Department of Mines' request, and Rex was transferred to Newfluor's payroll.[131] Presumably, the agreement did not stipulate that there were to be daily workplace readings because Rex continued Newfluor's practice of weekly readings, and the company did not appoint a second technician.[132] The government now had to send down its own technician periodically to check on Rex's readings.[133]

There seems to have been no protests at the failure to implement daily workplace radiation monitoring.

The practical merits of daily workplace readings were never discussed publicly. During the 1960s, alpha radiation measurements in the St Lawrence mines were the lowest reported from the worldwide mining industry (C13).[134] Furthermore, experience had shown that as long as the mechanical ventilation was functioning, there was no need to be concerned.[135] In such a stable radiation scenario, frequency of measurements would have to be a matter of opinion, but often enough to calculate plausible WLMs for workers. The Royal Commission presented figures on frequency of monitoring working sites in uranium mines in Canada and the United States. They ranged widely, the most frequent being Ontario, two to three times weekly.[136]

PROCESSING CLAIMS

When it came to the recommendations of the Royal Commission dealing with the processing of claims and compensation, some were accepted by the government while others were refused. The item that specified that if an autopsy had not been performed to substantiate a clinical suspicion of lung cancer, the family's claim for compensation "be given the benefit of the doubt" (R24) was rejected.[137] The recommendation concerning the constitution of an ad hoc committee of three physicians for dealing with appeals from claims rejected by Workmen's Compensation Board was accepted (R50).[138] The recommendation that regardless of the ceiling on a widow's pension from Workmen's Compensation Board, every dependent child should receive its stipulated allowance was accepted (R64).[139] The recommendation that widows on Workmen's Compensation be told of their entitlement for benefits from the Department of Welfare had already been implemented (R52).[140]

THE SPECIAL FUND

In principle, the government accepted the recommendation for a special fund to provide additional financial support for St Lawrence "disabled workmen" or their families in receipt of pensions from Workmen's Compensation (R53, R54).[141] Eventually, in 1973, the government set up the fund with an annual budget of $100,000, to run for ten years.[142] Besides the provincial government, the federal government, Alcan, and the Seibert family were each asked to make a commitment to contribute 25 per cent to the annual budget. The fund was to be administered by the Department of Welfare. Later Workmen's Compensation assumed responsibility for operating the fund.[143] A local committee was constituted to draw up the ground rules for expenditures from the fund.[144] The reason for the delay in setting it up was the failure of the Seibert family to respond to repeated requests for support.[145] The federal government also saw no need to contribute to the fund, leaving Alcan and the provincial government to provide equal proportions annually.[146]

To be eligible for support from the fund, a worker or his widow had to fulfill several stipulations. He or she had to be in receipt of a pension for an "industrial disease" from Workmen's Compensation. In 1969, the Workmen's Compensation Schedule of Industrial Diseases entitled to compensation listed only cancer of the lung due to exposure to radiation and pneumoconioses such as silicosis.[147] The family salary had to be no higher than $7,000 annually.[148] The head of the family or his widow was to receive a full share ($30.86 per month), the wife of a husband still alive received half a share ($15.43), and a dependent child received a quarter share ($7.72).[149] Benefits were made retroactive to July 1971.

The special fund has continued to this day with slight increases in the amounts of payments. As of 2009, a full share pays $36.63 per month.[150] As further diseases were added to the Workmen's Compensation Board Schedule of Industrial Diseases, the numbers entitled to benefits from the special fund also grew. Furthermore, at the initiative of the government, accidental workplace death was also made an acceptable qualification for access to the fund.[151] If entitling other conditions besides lung cancer for eligibility for access to the special fund is commendable on humanitarian grounds, its rationale is debatable. In the days before Workmen's Compensation, a common line of defence by operators who were sued by injured workers was to take their stand on one or more of "the unholy trinity of defence" – voluntary assumption of a known risk by a worker or negligence on the part of himself or a fellow worker.[152] The first of these could have been viewed as a rational exclusion criterion for deciding who should qualify for support from the special fund. Cancer of the lung in the St Lawrence fluorspar mines was a first ever event to be reported in non-uranium mines. The men of St Lawrence and nearby communities who signed up for work in the fluorspar mines had no idea that they were running a risk for developing cancer of the lung. There was every reason for treating such victims as special cases, perhaps outside the scope of Workmen's Compensation, as Judge H.A. Winter had recommended in his committee's report reviewing the operations of the Workmen's Compensation Act in 1966.[153] It was but just to compensate them more than Workmen's Compensation stipulated because of their involuntary assumption of an unknown risk. But could the same be said of the miners who developed silicosis? For them, the justification for granting access to the special fund was less strong than in the case of cancer of the lung. It is true it was widely believed that silicosis posed no risk because of the dampness of the mines.[154] Still, by the time of the trade dispute board in 1942, the union executive was well-aware of the risk of silicosis when it pressed Fraser to recommend mandatory x-ray surveillance of the workforce.[155]

ADDITIONAL COMPENSABLE CONDITIONS

The Royal Commission recommended adding a number of conditions to the Workmen's Compensation Schedule of Industrial Diseases.[156] The list included impaired pulmonary function and chronic obstructive lung disease in the pre-1960 workforce (R28, R21, R60). It had also recommended without qualification adding to the schedule tuberculosis, silicotuberculosis, and cancers of the mouth and upper respiratory tract (R41, R58, R62). The rationale for doing so was that radiation might have played a role in the genesis of these conditions. The government rejected these six recommendations except for silicotuberculosis.[157]

DISCRETIONARY CLAUSE

The Royal Commission recognized there should be a discretionary mechanism for compensating workers from the pre-1960 workforce who had already or would become unfit to work underground from conditions not listed in the Workmen's Commission Board Schedule of Industrial Diseases. In R19 and R20, it set out the procedures for compensating such cases.[158] The government accepted these two recommendations but modified them to require such claims, if from the 1951–60 workforce, to be referred to the Workmen's Compensation Board Committee of Medical Referees for settlement with the right of appeal to an ad hoc committee appointed as in R50.[159] Claims from workers who had taken their severance in 1951 or earlier were to be referred to the special fund.[160]

Smallwood's fears about the Royal Commission landing his government in trouble proved to be unfounded. It gave him a respite from the relentless criticisms he was facing. However, the Royal Commission did not put a stop to criticisms of Workmen's Compensation Board and the Aluminium Company of Canada. If anything, the criticisms became more strident.

7

Newfluor in the 1970s and Closure

By the time of the Royal Commission, people had begun to wonder how much longer Newfluor would continue to operate and whether it might not be prudent to think of alternative employment in St Lawrence. The brief submitted by the St Lawrence town council raised this last point and the Royal Commission encouraged the suggestion (R66).[1] The provincial government agreed it was a matter for the federal and provincial governments to attend to but did nothing further.[2]

Nevertheless, in the early 1970s Alcan appeared as committed as ever to developing its St Lawrence holdings. Blue Beach was reactivated. Since Blue Beach and Director were becoming exhausted, work on a new mine, Tarefare 2, meant to replace the other two, commenced at a cost of $10 million.[3]

ADDITIONAL COMPENSABLE DISEASES

Successive amendments to the Workmen's Compensation Act increased widows' pensions and dependent children's allowances as well as total disability pension rates (see Appendix F). In the early 1970s, the Moores government further expanded the number of diseases that qualified for admission to the Workmen's Compensation Board Schedule of Industrial Diseases. COLD and any cancers arising at any site in workers exposed to radiation were both admitted to the schedule.[4] The scientific basis for this step with respect to all cancers is not clear. The only known hazardous level of radiation to which the miners had been exposed was alpha. Because of its physical-chemical properties, the respiratory and possibly the gastrointestinal tracts would be the only parts of the body susceptible to its effects. In 1973, handling, in addition to extracting, fluorspar became grounds for compensating individuals developing any of the diseases listed in the Schedule of Industrial Diseases.[5]

In the past, while the risks from radioactivity in the mill houses of the Colorado Plateau uranium mines were not as high as underground, they were not negligible. Significant elevations of longer-lasting radioactivity, other than alpha, and increased rates of lymphatic and hemopoetic tumours were reported.[6] The levels of gamma and beta radiation in the St Lawrence fluorspar mines were not considered hazardous.[7] The physical-chemical properties of radon and its progeny would not have permitted their persistence in the mill houses. Consequently, radiation should not have been a problem at these locations. However, some workers had not worked exclusively on the surface or underground, but in both.[8]

ANGER TOWARDS WORKMEN'S COMPENSATION BOARD

For a number of reasons, the liberalization of entitlement to compensation, far from diminishing discontent directed at the Workmen's Compensation Board, only fuelled it. For example, a worker who put in a claim for COLD just before this diagnosis was added to the Workmen's Compensation Board Schedule of Industrial Diseases would have had his claim rejected. Later, when COLD had been added to the schedule, he would have been notified that his claim had now been accepted. He would then be tempted to boast that Workmen's Compensation Board had caved in to his unremitting pressure.[9]

Another cause of complaint was partial disability pensions.[10] Partial disability pensions worked reasonably smoothly when the Workmen's Compensation Board was compensating only accidents and injuries. However, when non-lethal diseases such as COLD were added to the list of compensable conditions, it was another story. More than anything else here, there was room for bitterness and misunderstanding. For example, a worker suffering from a disease such as COLD might be considered by Workmen's Compensation Board to be only partially disabled and still fit for light work. He would be awarded a partial disability pension, a percentage of a total disability pension (e.g., frequently 30 per cent). He was not expected to support himself on this pension but to be able to find a lighter job. His pension was compensation, nothing more. However, a worker receiving a partial disability pension for an accident, say for loss of a limb, was otherwise healthy and stood a good chance of finding and being able to perform a handicapped person's type of job. Not so for a person on a partial disability pension for an illness such as COLD; he was not otherwise healthy and could experience difficulty finding or being able to hold a job within the limits of his disability. Newfluor does appear to have tried to accommodate such cases in surface jobs, and, at the time of closure, had about a dozen lighter job employees on its payroll.[11] A partially disabled pensioner unable to find or hold a job within his capability was eligible to apply to welfare or the special fund for assistance.[12]

Once the ceiling on allowances for dependent children in the Workmen's Compensation Act was removed, social assistance benefits were at every level less generous than Workmen's Compensation. Naturally, unemployed, partially disabled pensioners agitated to be given support entirely from Workmen's Compensation Board, citing this as an entitlement, and to avoid the stigma of being dependent on welfare.[13]

In his insightful analysis of the operations of Workmen's Compensation in the St Lawrence experience, Elliott Leyton shows how incomprehensible and unacceptable the organization's workings were in the eyes of its clients in St Lawrence.[14] On the other side of the coin, the staff of Workmen's Compensation Board were equally frustrated at the inability or unwillingness of its clients to understand or accept the reasoning behind its rulings. The two parties, Leyton suggests, were living in different worlds, unable to understand or to communicate with each other.

Such a situation called for a redress. One suggestion was to treat the St Lawrence lung cancer epidemic as a special case outside the scope of the Workmen's Compensation Act.[15] A more radical proposal was to combine Workmen's Compensation and all the other forms of social assistance into one monolithic social assistance scheme from which to dispense assistance according to social and economic needs, regardless of cause.[16] Neither of these two suggestions was taken up by the government.

CRITICISMS OF ALCAN

A disastrous event like the radiation-induced epidemic of lung cancer at St Lawrence was bound to attract national attention. Ian Adams's article, "The Forgotten Miners," in *Maclean's* magazine in

1967 alerted the Canadian public about the tragic events unfolding at St Lawrence.[17] But it was perhaps Leyton's book, *Dying Hard: The Ravages of Industrial Carnage*, an oral history published in 1975, that did most to foster a critical view of the handling of the health crisis at St Lawrence. Alcan did not welcome Leyton's book, nor the spate of critical reviews its publication unleashed. One of which was quite inflammatory, recommending Leyton should have gone after "the real culprits, Seibert, Alcan and [Premier] Frank Moores."[18] Alcan made no attempt at a rebuttal to Leyton's book, choosing to believe that his strictures were not aimed at Alcan so much as at society's callous indifference to the suffering caused by industrial activities.[19]

Not all the opinions expressed publicly about Alcan were negative. There were voices raised in its favour. A disabled miner told Leyton the company "had treated him alright."[20] A widow of one of the miners who had succumbed to cancer of the lung had this to say: "People are always forgetting the many things Alcan did in St Lawrence. There has always been good input (*sic*) from the company. Whenever anybody needed a bulldozer or a front-end loader, Alcan always provided it."[21] Alcan also provided a financial support network for its employees, which included permanent disability pensions for workers who were not eligible for Workmen's Compensation, retirement income, and life assurance policies as well as illness and injury policies, all funded entirely by the company.[22]

Newfluor also brought many tangible benefits to the community. St Lawrence was the first town on the Burin Peninsula to have paved roads, running water, and a sewage system.[23] Alcan contributed $50,000 annually to the municipal council in lieu of taxes and $1,000 annually to each of the two schools.[24] Alcan employees paid taxes to the municipality – $20,000 in 1977 – which constitutes an indirect company contribution to St Lawrence.

SOCIAL EFFECTS OF THE EPIDEMIC

A catastrophe that killed so many men in their prime was bound to have a devastating impact on a traditional patriarchal society where the husband was the head of the family and the provider.[25] In 1968, with a population of 2,500 and 371 households, seventy-seven (20 per cent) were fatherless due either to lung cancer (fifty-five) or silicosis (twelve).[26] The death rate for young males was likened to what would be expected in warfare. These statistics told part of the story. The widows, another study found, were ill-equipped to step into their husbands' shoes.[27] The community experienced an increase in "illicit" relationships, more "out-of-wedlock" babies, ill-disciplined children who performed poorly at school, and finally teenage male delinquency.[28]

UNION MILITANCY

The early 1970s were also a period of increasing union militancy at Newfluor. Strikes were frequent. The stage for this militancy was set in 1962, when the St Lawrence Workers Protective Union requested and was granted affiliation with the Confederation of National Trade Unions (CNTU).[29] This affiliation brought benefits to the local union, such as skills and greater clout in negotiations as well as the financial resources to maintain long strikes. Still, there was a price to pay for these benefits. The parent organization's goals and those of a local affiliate did not always coincide. Furthermore, the union at Arvida had parted ways with the CNTU and affiliated instead with the Fédération des syndicats du secteur de l'aluminium.[30] As a result, the unions at Arvida and St Lawrence appear to not have coordinated their actions. Additionally, the local affiliate could be badly advised by the national office. There was the suspicion that the CNTU was egging on the St Lawrence Workers Protective Union, encouraging demands for unrealistic wage increases with assurances that, as the sole provider of fluorspar to Arvida, it had a stranglehold on Alcan.[31] The pursuit of wage parity with Arvida and Kitimat, BC (the other Alcan smelter) was defended on the grounds of the higher cost of living at St Lawrence and mining being a more dangerous occupation than smelting.[32] Perhaps the major reason for labour unrest in the 1970s was that these were the years of galloping inflation. Unions negotiated what appeared to be a generous wage package only to see it evaporate due to rising prices.[33] Between 1971 and 1975, there were three strikes and then a lockout – the last, the longest-lasting of the industrial actions in the history of the fluorspar mines.[34] In the eyes of many in the community, this lockout was the decisive event that propelled Alcan down the road leading to closure of its St Lawrence holdings. On 31 March 1975, the collective agreement between the union and Alcan expired.[35] Negotiations for a new agreement broke off on 18 April. On 27 May, the first ship of the season arrived to start loading the winter stockpile of fluorspar but was prevented from doing so by a picket of miners' wives.[36] The company waited a few days and then closed down operations on 8 June. It insisted this was not a lockout.

DYER REPORT I

Attempts on the part of the government to bring about a settlement failed. Then, in compliance with the Labour Relations Act, it appointed Howard J. Dyer, a professor of mechanical engineering at Memorial University of Newfoundland, as a one-man commission to inquire into the dispute and to make recommendations "at creating general labour peace in the industry."[37] His terms of reference stipulated, firstly, that he was to resolve the impasse over wages and then, at a more leisurely pace, deal with safety in the mines. How far apart the two sides were on wages was shown in the almost twofold difference in the wage packages each put on the table. To support its case, the union presented a brief prepared by the Economic Research Department of the Confederation of National Trade Unions, purporting to show that in terms of real wages, Alcan's offer would put workers in a poorer position in 1977 than in 1973, assuming cost of living would continue to rise at the 1975 rates.[38] The union also had to defend its wage demands from the charge that they would make St Lawrence fluorspar even less competitive on the open market. The brief dismissed this concern by citing the shortages of fluorspar since the late 1960s and, consequently, rising prices for it on the world market. Still, the union must have had some unease about being so confident that its wage demands would not reduce Newfluor's competitiveness, for when it came to the ultimate wage settlement, the union indicated it would be flexible, provided real earnings would be higher in 1977 compared to 1975.

Dyer submitted the first part of his report to the government on 18 August 1975.[39] He questioned several of the assumptions behind the wage package in the union's brief. The "wage erosion" picture the brief painted was more gloomy than was warranted because it had not factored in the company's proposed cost-of-living allowance. When it came to complaints concerning wage disparities between St Lawrence and Alcan's mainland operations, Arvida and Kitimat, the complaints could only be substantiated by comparing wages for the same or similar jobs at the three plants. Finally, the brief's rosy picture of a continued shortage of fluorspar, and therefore high prices, could turn out to be mistaken.

The commissioner had to come up with a wage package that stood a good chance of acceptance by both parties or could serve as a basis for a negotiated settlement. He pointed out that Alcan's wage offer was not unreasonable and was higher than the wages at the other mining companies in the province. (For comparison purposes, in wage negotiations it is now customary to present wages also in relative terms, i.e., as percentages of national averages.[40]) However, Alcan's offer was not acceptable to the union. Consequently, in spite of his reservations, the commissioner had to come up with something higher. Here he ran into an insurmountable problem – the high cost of producing St Lawrence fluorspar regardless of wages. This was the same constraint that had forced the closure of the Corporation over a decade earlier. In 1974, Alcan was paying eighty-nine dollars per ton upon delivery of Mexican acid grade compared to an estimated $110 per ton for St Lawrence fluorspar, brought up to acid grade at Arvida.[41] On the basis of Alcan's wage offer, Dyer calculated the company would be paying twenty-one dollars per ton more by using St Lawrence rather than Mexican fluorspar. The reasons Alcan put up with paying more than it had to for fluorspar, besides the need of an assured supply, was the fear that should it close the St Lawrence mines, world prices might rise to a level high enough to offset the savings by dispensing with St Lawrence fluorspar. Dyer was sceptical about this possibility and wondered how much longer Alcan would go on paying an exorbitant price for St Lawrence fluorspar. He could have added that the unsettled labour relations climate at St Lawrence in the 1970s meant that it could no longer be counted upon as an assured source for fluorspar. On the other hand, Mexico, which had seen coups and uprisings in the past, had been politically stable since the mid-1930s.

The wage package Dyer came up with was close to Alcan's offer – an extra one dollar per hour for the first year and an additional forty cents per hour for the second. He admitted his wage package could not be justified on economic grounds but hoped both parties would accept it. Dyer made several observations that can only be interpreted as veiled cautions to the union. Alcan, he informed the union, had purchased enough fluorspar on the open market to provide for its needs until the autumn and had taken out an option to purchase a further supply to tide it over the winter. More alarmingly,

M.E. Gooding, the plant manager at Newfluor had revealed as far back as 1971 that Alcan had purchased fluorspar "holdings" in Mexico.[42] Alcan rejected Dyer's wage package.[43]

DYER REPORT II

After he finished with the wage question at Newfluor, Dyer turned his attention to safety in the mines.[44] The mines had a high accident frequency rate – 15.8 per cent higher for the years 1961–75 than the average of all the other mines in the province. This rate has to be squared with the report of the Royal Commission, which showed that, for the years 1957–67, Newfluor's accident frequency rate was usually below those of the other mines in the province.[45] However, an examination of Dyer's and the Royal Commission's accident frequency tables show that Newfluor's poor performance in accident frequency rates began in 1969, after the Royal Commission, and had continued year by year up to 1974. There is no obvious explanation for this alarming trend. Dyer observed that almost half of the accidents were workers with less than one year's seniority and those mostly had less than six months. (There had been an increased worker turnover rate following the discovery of high radiation levels in 1959.[46]) This pointed, so Dyer said, to the need for more effective indoctrination of new recruits into safe working practices.

On the bright side, dust measurements in the mill were one-third of permissible levels. Dyer's mandate did not call for an in-depth study of radiation safety in the fluorspar mines. However, he noted weekly readings (WL) at seventy to eighty locations, ranging from 0.20 to 0.30 WL and considered to be safe. If a reading rose to 0.5 WL, the ventilation system was checked and at 1 WL the work area was evacuated. Readings were posted on the miner's bulletin board. Copies were sent to the interested parties and the Department of Mines. Dyer's report included all locational readings (WL) for one month, and one worker's personal accumulated radiation exposure (WLM) for the same month. They were well below permissible levels. Since 1968, the highest accumulation for a worker had been 11.85 WLM (1,000 WLM was considered to be the maximum safe accumulation dose). There had been only one case of lung cancer among miners hired after 1960, and he had worked only three years underground. The peak of the lung cancer epidemic had passed, and only the occasional case was turning up. Dyer quoted Reuben Yourt,

P Eng, an engineer who had done consulting work for the government in the mines, as stressing the urgency of persuading the miners to stop smoking and thereby reducing their risk of developing lung cancer by ten to twenty times. The ventilation system consisted of four fans at Director mine, two at Tarefare mine, and one at Blue Beach mine. If a fan stopped, an alarm sounded immediately. However, when Dyer visited St Lawrence, the company's air sampling pump for radiation monitoring was not functioning, which left only the Department of Mines' pump for taking readings. Replacement of the company's pump was being delayed by failure of the Department of Mines to advise Alcan what model to purchase. Dyer emphasized the urgency of getting additional pumps quickly.

Dyer also addressed the problem of rock bursts, which were an ongoing serious hazard. To some extent, the risk was related to mining methods. With the expert advice of Professor R.G.K. Morrison, Newfluor had been changing over to "Longwall" sequential mining, which the mining staff were convinced was reducing both the frequency and severity of rock bursts.

The suspension of operations dragged on. At one point, Alcan, according to the recollections of one of its management team, came up with an offer that most of the local union members wanted to accept but were talked out of it by the Confederation of National Trade Unions representative telling them they had Alcan at their mercy. The government continued its efforts at mediation. Eventually, agreement was reached and employees returned to work 9 February 1976.[47] The suspension of operations had lasted seven months. Things did not return to the way they had been before the suspension of operations.

DOWNSIZING

When Newfluor recommenced operations in February 1976, it was a shadow of its former self. Its payroll was down to two hundred employees, half of what it had been there before the suspension of operations, and working only one shift.[48] This drastic curtailment of operations was imposed because during the strike Alcan had been purchasing Mexican fluorspar. The stockpile that had accumulated during the winter of 1974–75 was still on the dock at St Lawrence, waiting to be moved up to Arvida. Assuming that Alcan stopped purchasing Mexican fluorspar, there was no urgent need to get the

St Lawrence mines up to pre-lockout levels of output until the stockpile had been exhausted.[49]

Alcan seemed as committed as ever to using St Lawrence fluorspar. They spent three million dollars to build a new mill.[50] Work was restarted on sinking a shaft for Tarefare 2, and, by 1977, $5 million of the estimated cost of $10 million had been expended on this project.[51] Tarefare 2 was to replace Director and Blue Beach South mines, which were becoming exhausted and structurally obsolete. Middle management was told production had to become more efficient and the gap narrowed between the cost of mining, shipping, and upgrading St Lawrence fluorspar to acid grade and the delivery price of Mexican fluorspar.[52] Middle management responded to the challenge. With only one shift, it was able to bring output up almost to pre-lockout levels. At the same time, costs were brought down almost to the level suggested by head office to aim for. Unfortunately, every time the price of Mexican fluorspar dropped, a further reduction in costs from St Lawrence was demanded by Alcan.[53] Furthermore, to complicate matters, shortly after Alcan recommenced operations in St Lawrence, it was faced with a strike in Arvida lasting from June to November 1976.[54]

NEWFLUOR CLOSES

By 1976, then, there were no practical or economic reasons for Alcan to continue mining St Lawrence fluorspar apart from recouping some of the money it had invested in its operations. However, once the suspension of operations had ended, work had begun once again on getting Tarefare 2 into production. It must therefore have come as a surprise when Alcan announced its intention to close its mines in St Lawrence.

Alcan's decision to close its St Lawrence operations was communicated privately to the government, and then, a month later, at the end of June 1977, made public. The government was able to get a three-month stay of closure to 1 February 1978.[55] Alcan announced that it took this step "with extreme reluctance and great regret" after an internal review of its operations showed that they could not be continued economically.[56] Alcan told the premier that it had been set back $3 million in 1976 by its use of St Lawrence rather than Mexican fluorspar.[57] The only reason it had not taken the step sooner was because of worldwide shortages of fluorspar in the early seventies.[58]

HODGE REPORT

Upon receipt of notification of Alcan's intentions, as a first step the federal and provincial governments jointly hired B.L. Hodge and Partners, a UK consultant firm, to carry out an assessment of Newfluor's mining operations to determine if they were or could be made viable and, if so, what measures would be required to achieve this goal. In his report submitted in November 1977, Hodge concurred with Alcan that the existing operations were no longer viable because of an "interrelated combination of high costs and low productivity ... [requiring] the introduction of economies and improvement in efficiency."[59] Specifically, Newfluor's mining methods were inefficient, the ratio of management/supervisory staff to workers was too high, and miner output was suboptimal. The veins being worked were only about 43% CaF_2. This ore was subjected to beneficiation to bring it up to around 66% CaF_2 and 24% SiO_2, before being shipped to Arvida for further beneficiation up to around 97% CaF_2 and within a range of 3–4% SiO_2. This was wasteful. The entire beneficiation process, Hodge said, should be carried out at St Lawrence, and the product then shipped to Arvida as filter cake with no more than 10% H_2O, 1.5% SiO_2, and 97% CaF_2, for immediate use in smelting. This would require renovation of the existing heavy media separation plant. The Director mine was obsolete and should be closed, leaving Tarefare 1, Blue Beach South, and the as yet to be opened Tarefare 2 to be the producers of fluorspar. When it came to markets, Europe was self-sufficient in fluorspar. In the United States, St Lawrence would have to compete with foreign fluorspar. This left only Canada. Because of the high silica content, St Lawrence fluorspar was no longer desirable for steel manufacturing. Therefore, the acid grade market was the only one open to sales of St Lawrence fluorspar. To produce this grade at a viable price would require $9.5 million to upgrade the mill and other facilities. Were the St Lawrence mines to be sold, a further purchase price of $3.5 million would have to be added on to this estimate. Hodge was sceptical that, even with the streamlining and changes he recommended, it would be possible to produce a product that could compete on the open market.

Hodge had nothing to say about the unsettled labour climate at St Lawrence, apart from a brief reference that it might have been one of the reasons Alcan was leaving. Nor did he mention the high electricity bills from keeping pumps and ventilation equipment running. Hodge's report contained conclusions and no recommendations. In his summary of the most important conclusions, he had this to say: "The decision by Alcan to terminate captive production ... is questionable" because the current surplus and low prices were unlikely to continue.

Hodge expressed the hope that by rationalizing its operations, Alcan could be induced to stay on, but, failing this, the company should be asked to leave the mines and the installations in good condition for sale. He did not mention the point that the purchaser would probably have to operate an independent mine selling on the world market. Yet, he was sceptical of the feasibility of producing fluorspar at St Lawrence at competitive prices due to the high cost of removing silica.

In the days that followed the announcement of impending closure, Alcan made it clear why it was taking this step.[60] It stressed the disadvantages of St Lawrence fluorspar compared to Mexican fluorspar. Furthermore, there were rock bursts, necessitating shutting down operations for days at a time, and, finally, the high cost of keeping the mines dry. Alcan had acted only after it had conducted an internal review, which had demonstrated, to its satisfaction, that fluorspar could no longer be mined profitably at St Lawrence.[61] Indeed, as Alcan's public relations officer, Bud Rudd, put it bluntly, it would take a miracle "to make a single cent" out of St Lawrence.[62] In other words, St Lawrence fluorspar had always been expensive to produce; now technological advances in smelting had administered the coup de grâce to its industrial usefulness. Still there were suspicions that the unsettled labour situation and the adverse publicity from the cancer of the lung epidemic were also unspoken reasons for ending mining St Lawrence fluorspar.

If mining fluorspar at St Lawrence was so costly, Newfluor's inefficient mining methods, as Hodge charged, must have been partly to blame. Alcan was not a mining company. It had other interests besides operating a mine. It was with these interests that Newfluor competed for attention and funds. A member of its middle management team, while praising Alcan as the best company for which he had ever worked, admitted there were times he found head office's lack of mining expertise to be frustrating.

RESPONSES TO CLOSURE

From the point of view of its public image, Alcan's announcement of closure could not have come at a more inopportune time. Alcan had just made a record profit of $85.2 million for the first half of 1977.[63] The $3 million that Alcan claimed it had lost in 1976 by using St Lawrence rather than Mexican fluorspar, critics said, was not a loss but a decrease in profits.[64]

In Newfoundland, the closure of an industry, the sole provider of employment in a community, was bound to have political repercussions. Federal and provincial politicians of every hue united in denouncing the decision: "Morally repulsive" and "criminal" were some of the terms thrown at the company's decision.[65] Another critic called for penalties, "no concessions, no help in any way."[66] Alcan was accused of sacrificing St Lawrence to make a larger profit from "the blood and sweat of Mexican miners."[67]

While it was awaiting Hodge's report, the provincial government asked the federal government to impose a tariff on imported fluorspar.[68] The request was turned down, presumably for the same reasons it had been refused to the Corporation twenty years earlier.[69]

Upon receipt of the Hodge report in November 1977, Brian Peckford, the provincial minister of mines, urged Alcan to reconsider its decision.[70] Alcan refused. Having spent $10 million improving its St Lawrence holdings in the years 1971–75, it was not prepared to spend more.[71]

ASSETS DISPOSAL

With no interested purchasers in the offing, all that remained was to negotiate the terms for the disposal of Alcan's assets in St Lawrence. Alcan offered a five-year lease of its St Lawrence installations and mining rights at one dollar a year to the provincial government.[72] This would give the government time to find a purchaser, but would have required $200,000–$300,000 a year to maintain the properties.[73] The government rejected this offer.[74] Instead, it demanded that Alcan transfer ownership of its facilities to the government for one dollar.[75] This in turn was rejected by Alcan, which cited $10 million as the value of its holdings in St Lawrence, and noted that it would be willing to sell its assets and mineral rights for this figure.[76] The government countered by offering Alcan retention of its mineral rights in return for maintaining the mines and surface buildings in an acceptable condition.[77] Mothballing was evidently not acceptable to Alcan, and with it ended all hopes of a negotiated settlement. The government was in no mood to be easy on Alcan, and Peckford had warned that if agreement could not be reached, the government might expropriate Alcan's St Lawrence holdings.[78]

By the terms of the Mines Act, a lessee was required to remove all equipment and dismantle all buildings within six months of abandoning a lease.[79] With the threat of expropriation hanging over its head, Alcan moved quickly. The company dismantled and sold all its movable assets in St Lawrence to an American company.[80] The government retaliated by amending the Crown Lands Act, lowering the reversion of mineral rights to the Crown to five years after a lease was terminated or abandoned.[81] (Alcan's mineral rights had been held in perpetuity by agreement with the government.[82])

FINANCIAL MATTERS

Three matters had to be settled to complete Alcan's winding up its affairs in St Lawrence. The first of these was severance pay. Unionized employees received one week of severance pay for every year of service.[83] Management and clerical staff, as well as those laid off a year earlier, received more generous settlements.[84] Alcan also stated its intention of continuing to pay the insurance premiums of those hired before 1960 but made no commitment to the 30 per cent partially disabled workers employed in surface jobs.[85] The second item was the special fund. Alcan promised to continue its share annually to the fund for another two-year period to 1981 and then, subject to review, possibly for another two years.[86] The size of the fund depended on the number of claims. It peaked at $120,000 in the early 1980s but had dropped to $80,000 by 1986.[87] Alcan ceased its contributions to the fund in 1985, leaving the Newfoundland government to bear the full costs.[88] The final matter was what to do with the remaining, about one hundred, workers who were still around at the end of 1977.[89] Alcan offered jobs to some at its plant in Kitimat, BC and considered sending in recruiters from other mining companies.[90] Most of the miners were reluctant to move, which prompted David Culver, the president of Alcan, to say, "the world doesn't owe you a job where you happen to want to live."[91] Unlike when Wabana closed down, the government did not provide mobility grants or upgrading and retraining programs.[92] Presumably, while the closure of Wabana was accepted as final, that of the St Lawrence mines was viewed as temporary; the new owners were expected to hire the former Newfluor workers.

The municipality of St Lawrence greeted the news of the mines closure with dismay. No longer could it count on the grant from Alcan to underwrite the town services.[93] These would now have to be cut back and plans for a shopping mall shelved.

The end of Newfluor provoked post-mortems. What should have been done to prepare for that day? One suggestion was that, in the future, mining companies should be required to pay disturbance fees like the petroleum companies operating off the Shetland Islands where the fees went into a fund to develop alternative industries when the petroleum fields were exhausted.[94]

In spite of Alcan's gloomy verdict on the chances of mining St Lawrence fluorspar profitably, the government refused to give up on finding a company willing to take this chance.

8

St Lawrence Fluorspar Company Ltd

The federal and provincial governments were both committed to fostering economic development in the Burin Peninsula. To this end they had established the Burin Peninsula Development Fund, on a federal-provincial cost-shared basis (70:30), to provide financial assistance to companies willing to start up in the region.[1] Reactivating the fluorspar mines at St Lawrence seemed to the provincial government one project worth pursuing, even though the Hodge report had given no countenance to the viability of an independent fluorspar mine in St Lawrence selling on the open market.[2] It could do nothing until 1983, the year Alcan's mineral rights reverted to the Crown. In the meanwhile, the government was under pressure to find alternative forms of employment. A fish plant was opened in 1978, only to close in 1982.[3]

THE NEW COMPANY

In 1983, in response to international advertising, the government found a company interested in reactivating fluorspar mining at St Lawrence, called Minworth PLC, a UK company, which owned and operated fluorspar and barite mines in the UK.[4] Minworth formed a subsidiary, the St Lawrence Fluorspar Company Ltd to operate its St Lawrence holdings.[5]

The plan was to operate two underground mines, Blue Beach North and South, and several open pits.[6] The ore would be brought up to acid grade level (97% CaF_2) with less than 1% silica, for sale on the world market.[7] The world demand for fluorspar was still depressed, but it was anticipated it would soon recover and create a shortage and a rise in prices.[8] The new company would be starting from scratch since all Alcan's surface installations had been removed. Reactivating or opening new mines and constructing a new mill and other buildings would take time and money – an estimated $14.4 million.[9] The St Lawrence Fluorspar Company accepted an offer of $6.8 million from the Burin Development Fund.[10] The rest of the financing, it was announced, would be raised by partnership with a private unnamed Canadian investor.[11] In the meanwhile, the province provided $1.5 interim financing.[12] One reason for the federal government's support of such an uncertain financial venture was strategic considerations. Just as during World War II, St Lawrence would be the only large-scale source for fluorspar in Canada. The project would create about one hundred new jobs.[13]

NO STRIKES COMMITMENT

An essential ingredient for success of the project would be stable labour relations. Since it was widely believed in St Lawrence that the strikes and lockout in the 1970s had been one reason for Alcan pulling out, the community was not going to risk a repeat of this mistake.[14] An informal understanding was reached between the company and the St Lawrence municipal council that workers would not go on strike or join a national or international federation of trade unions.[15] Minworth claimed it had just suggested, not insisted on, these terms because under the Labour Relations Act it was unlawful to prohibit workers from forming a union.[16] The Newfoundland Federation of Labour Unions denounced this arrangement, claiming the wages workers would receive would be 40 per cent lower than those paid to miners in Ontario.[17] The mayor of St Lawrence, who was also the plant manager, responded by saying it was none of the federation's business.[18] The prospective miners seem to have been quite satisfied to abide by the understanding. Both parties realized the understanding could only be a temporary measure. Later, the workers were unionized and, unlike the previous union, affiliated with the United Steel Workers of America, instead of the Confederation of National Trade Unions.[19] Wages became comparable to those paid in the other mines in the province.[20] By the end of 1986, Blue Beach North mine had been opened and the mill constructed.[21] The federal government had repaired the wharf to the tune of $1.5 million.[22] The first shipment of fluorspar left in 1987.[23]

Unfortunately, it did not meet the purchasers' specifications for particle size (it was ground too fine), and the grinding machinery had to be reset to produce a coarser grain.

Throughout its life, the St Lawrence Fluorspar Company Ltd was beset by shortages of capital. Negotiations with the unnamed Canadian partner, on which such high hopes had been set, came to nothing.[24] Consequently, the company was forced to look to the government and the parent company for financial assistance. In May 1987, to avert a layoff, the company took out a loan of $800,000 guaranteed by the province.[25] Later in the year, $3.8 million was infused into the company – $1.24 million from the province, $500,000 from Minworth, and the remainder in the form of guaranteed loans and deferred payments on a Barclay's bank loan.[26]

PRODUCTION FIGURES

Hodge's prediction of a turnabout in the world demand for fluorspar was unfounded. The lack of demand for fluorspar had a crippling effect on the operations of St Lawrence Fluorspar Company Ltd. Orders were slow in coming, which caused slowdowns or shutdowns until stockpiles could be sold.[27] The mill operated at 3.5 days a week when an order had to be filled.[28] There were times of complete shutdowns.[29] During 1987, there were four shipments of acid grade fluorspar, three to the United States and one to the UK.[30] In the next three years, there were a few more shipments and then none (see Appendix D). With a company like the St Lawrence Fluorspar Company Ltd, perpetually strapped for cash, there would be the fear that health and safety services might not receive the attention and investment they deserved. It was left to the Division of Occupational Health and Safety to deal with these very real concerns.[31]

SURVEILLANCE

When it was announced that the St Lawrence fluorspar mines were to be reactivated, there were dire warnings that the workers might be exposed to hazardous radiation levels.[32] A more practical point was the eligibility of former underground St Lawrence miners who had accumulated high amounts of radiation. The decision reached was that they qualified for surface jobs that did not involve trips underground.[33]

The division of Occupational Health and Safety also had its concerns.[34] Since the company was constantly strapped for funds and there were fears that there would be insufficient investments to maintain occupational health and safety services at an acceptable level – fears that unfortunately were to be fulfilled – requests to correct problems were ignored. Closure was not a political option. All the division could do was try to persuade or cajole the company into complying with its regulations. It became an annual event for the plant manager to be called into the division when he passed through St John's en route to the UK and to present him with a list of items needing attention, such as appointment of a safety inspector. But these efforts were to no avail. The government inspector, during his tours to monitor radiation levels, noticed that the union representative never voiced any concerns.[35] He wondered if this was due to fears that if he did the mine would close down and he and others would be out of work. Indeed, unlike the St Lawrence Workers Protective Union, which was constantly, vociferously complaining about health and safety at Newfluor, the St Lawrence Fluorspar Company Ltd union was strangely silent in this respect, probably for the reason stated above.

Although the St Lawrence Fluorspar Company employed a radiation technician, the community seems to have become uneasy about the radiation safety of their men folk in the mines.[36] In 1988, the Town Council was assured by G.K. Bradbury, chief inspector of mines, that it need have no concerns.[37] Shortly afterward, an internal letter from Bradbury to R.K. Langdon, the assistant deputy minister of labour, came into the hands of the press.[38] It painted a scandalous picture of the obstructions he encountered in his attempts to monitor health and safety conditions in all the mines of the province. Bradbury confessed that his assurances to the St Lawrence Town Council had been more confident than were warranted. During the past year, after continuous pestering, he had finally been able to obtain a few radiation readings from the St Lawrence mines, but no diesel exhaust fume measurements. Ventilation had been inadequate in the past but there had been no follow-up checks to see if this had been corrected. Bradbury urged Langdon to station an inspector full-time at the St Lawrence mines to take repeated radiation and diesel fume measurements as well as to check ventilation systems to ensure they were functioning. What Bradbury was advocating was verification of the company's radiation technician's readings, a mission similar to the one David Rex fulfilled at Newfluor from 1967 to 1970. Whether this letter spurred any actions is not recorded (I did not come across any rebuttals to Bradbury's charges) but when the press approached Bradbury some months later, he refused to comment on "health conditions" at the St Lawrence mines.[39]

CLOSURE

In January 1991, to the relief of the Division of Occupational Health and Safety, the St Lawrence Fluorspar Company Ltd was declared bankrupt.[40] The cause was attributed to depressed market conditions, the result of "aggressive marketing of Chinese fluorspar" that pushed down world prices.[41] The government cancelled the St Lawrence Fluorspar Company Ltd's mineral rights and began to look for a purchaser. The Department of Employment and Labour Relations provided a grant of $35,000 to maintain the mine site and its structures while the government continued to look for a purchaser.[42] As none were forthcoming, the government sold the mine and its buildings to the Greater Lamaline Area Development Association for one dollar.[43] Ernst and Young, who were appointed as receivers and managers of the St Lawrence Fluorspar's assets, considered the prospects for mining fluorspar profitably to be "uncertain."[44] A local group, Burin Minerals Ltd, a subsidiary of Burin Fluorspar Company, continued to promote the worth of St Lawrence fluorspar with the hope of attracting a company to exploit it commercially.[45]

How realistic were such hopes? On the plus side, in 1977 Hodge estimated fluorspar reserves to be just over eight million tonnes, enough at the then rate of extraction to keep the mines running for years.[46] Yet, three companies have abandoned fluorspar mining at St Lawrence, citing high costs of production and the high content of silica in the ore. But besides these drawbacks, fluorspar mining at St Lawrence had to contend with others inevitable in hardrock mining. Of all the different types of exploitation of non-renewable natural resources, hardrock mining is the most wasteful and destructive to the environment.[47] Disposal of waste rock and restoring the environment, especially if it has been contaminated with toxic wastes, can be costly. Then fluorspar, like other minerals, has to contend with wide swings in the price it fetches on world markets, depending on whether there is a glut or a scarcity. Hardrock mining is environmentally destructive and financially risky; it is also the least profitable of the enterprises engaged in exploiting non-renewable natural resources.[48] However, minerals are vital to the survival of modern societies. Industries that need them will shop in the cheapest markets. These are often to be found in less developed countries where wages and costs are lower.

Twenty years went by and, while there were inquiries from interested parties about reopening the mines, nothing came of them. In an early draft of this book, I originally ended on a pessimistic note about the chances of fluorspar mining ever returning to St Lawrence.

However, given the considerable reserves of fluorspar in St Lawrence, as well as advances in technology for removing silica and harmful substances from fluorspar concentrates, it still seemed like it was only a matter of time before a company that was willing to take a chance on mining fluorspar at St Lawrence would come along. Indeed, in August 2011, it was announced that Canada Fluorspar, in partnership with Arkema, a French chemical company, would be reactivating fluorspar mining at St Lawrence.[49]

Since the radiation disaster and epidemic of lung cancer loom so largely in the history of the St Lawrence fluorspar mines, they should be given the last word. What conclusions can be drawn? The epidemic, which began in the middle of the last century, must have almost run its course. What is its most recent health status? This and other epidemiological aspects of the health of the workforce require detailed discussion.

SUMMARY OF UPDATES ON A COHORT OF ST LAWRENCE FLUORSPAR UNDERGROUND AND SURFACE WORKERS

In 1988, and again in 1998 and 2007, updates were published on the most recent health status of a cohort of underground and surface workers (millers), employed in the mines between 1933 and 1978.[50] The original cohort included all workers in these two categories. The final cohort, analyzed in the first (1988) update, included only those subjects with adequate personal information. In the subsequent updates, the numbers of participants have fluctuated slightly because of changes in status of subjects due to new information.[51] By 2001, the year of the most recent health assessment, the size of the cohort was 1,742 underground workers and 328 millers (published 2007).[52] The most recent prevalence of the important diseases in this cohort and the role of alpha radiation are discussed in the rest of this section.

In the underground workforce, there were increased mortality rates for cancer of the lung and bronchi (191 cases), silicosis (7), infective diseases including tuberculosis (19), accidents (74),[53] chronic bronchitis, emphysema, and asthma (23).[54] Mortality rates were lower than expected for other diseases – possibly a manifestation of "the healthy worker effect." This principle states that workers in occupations requiring physical fitness and strength have a lower incidence of non-occupational diseases than the general population. There may be perhaps another principle at work in this workforce – "the principle of inequality."[55] It states that, for some unexplained reason, in any society there is an inverse correlation between incomes and morbidity rates. On any scale of incomes, miners would be regarded as being relatively well paid and should be relatively healthy. Miners though are at risk for developing occupational diseases. To what extent these two principles interact in this situation is a subject of debate.[56]

There were fifteen deaths from lung cancer among the millers exposed only to background radiation.[57] Among the underground workers, rates for lung cancer, as expected, correlated with cumulative exposure to radon and radon progeny.[58] More significantly, risk for lung cancer increased with duration of exposure.[59] That is to say, for two equal cumulative exposures to radon and radon progeny, the one reached sooner because of higher levels of radon and radon progeny and more rapid exposure rates posed less of a risk than the one reached later because of lower levels and slower rates. Others have observed the same phenomenon but Villeneuve and co-authors argued that it might be of no significance.[60] Still, it raises concerns about the safety of the low dose radon and radon progeny levels to which the St Lawrence underground workers hired after 1960 were exposed – the year enhanced ventilation was installed. In the most recent update, there were twenty-eight cases of lung cancer among the 743 miners hired after 1960.[61] The number is small. Even though in this group there was a correlation between risk and cumulative radon exposure, statistically it was not strong.[62] Investigators have suggested that with the passage of time additional numbers might increase the power of statistical analyses of this group.[63]

How important was the contribution of residential exposure to cumulative alpha radiation exposure in the St Lawrence group of miners with lung cancer? The St Lawrence workers were practically all recruited locally and had spent all their lives in that or neighbouring communities. All that was available on residential exposures to alpha radiation was some measurements taken of a sample of households.[64] The geometric mean WL was 1.7, similar to baseline residential levels in other Canadian communities.[65] The conclusion was that residential exposure to alpha radiation had not been a significant contributor to risk for developing lung cancer in the St Lawrence cohort.[66]

In any epidemic of occupational lung cancer, the contribution of cofactors has to be considered. In the St Lawrence cohort there were three known possible cofactors: smoking, silica, and diesel fumes. In the past, the St Lawrence miners were almost to-a-man heavy smokers.[67] There must be a suspicion that smoking contributed to the epidemic. In the second update, a relationship between smoking and risk for lung cancer was demonstrated.[68] However, in the third update, because of a new survey on smoking habits, no relationship was apparent. In this report, there was no statistical difference between "ever" and "never" smokers.[69] But there was an association between the magnitude of risk and the daily number of cigarettes smoked. The authors cautioned that these estimates were derived from only half the cohort, and the small number of cancer deaths in each smoking category handicapped analyses.[70] Inhaled silica is now recognized as a carcinogen.[71] While silicosis was present in the St Lawrence cohort, its prevalence was unknown and consequently its contribution could not be assessed.[72] The same is true of diesel fumes.

Was there a preponderance of a particular cell type in radon and radon progeny-induced lung cancer, and did it vary with cumulative radon and radon progeny exposure? Histological material for examination was available on only ninety-seven of the 191 lung cancer cases (slightly more than half).[73] Furthermore, only half of these could be classified as adenocarcinoma, squamous cell, or small cell undifferentiated – the three main types of lung cancer. These small numbers hampered meaningful analysis. However, the majority were squamous cell – the same type that was found with extended follow-up in the Colorado Plateau uranium miners and in Wright's group of cases of lung cancer from the St Lawrence fluorspar mines.[74]

Did the radiation hazard at St Lawrence predispose the underground workforce workers to other cancers besides lung cancers? There were elevated rates of cancers of the pharynx/buccal cavity,[75] the salivary glands, and the urinary tract.[76] Pooled analyses of this and other cohorts showed no statistical association between these elevations and cumulative radon and radon progeny exposure.[77] It was concluded that there must be some other explanation for elevated rates of tumours at these sites.[78]

Perhaps the 1988 update had one immediate practical effect. In 1988, the International Commission on Radiological Protection proposed raising radon and radon progeny exposure limits from 4.0 to 4.8 WLMs per year. Citing statistics on the St Lawrence cohort, Morrison warned this would lead to significant increases in lung cancer rates among miners exposed to radon and radon progeny.[79]

These three updates illustrate the handicaps retrospective (historical) cohort studies sometimes encounter, such as small numbers, lack of personal information, and new data requiring revision of previously held opinions.

It is on this somewhat uncertain note that the history of epidemic of radiation-induced lung cancer at St Lawrence has to be left. Almost eighty years have passed since the *Renunga* steamed up the bay, carrying in its hold the second-hand machinery that was to usher in the transformation of St Lawrence into a mining town, but bringing with it a mixed bag of benefits and tragedy. A new chapter in the life of St Lawrence is about to open, which makes this the appropriate point to end this part of its history.

Epilogue

In the past, mining fluorspar at St Lawrence has faced two handicaps that were never surmountable. One was the high cost of extracting and shipping the fluorspar. Alcan, the owner of Newfluor, its subsidiary in St Lawrence, accepted this state of affairs as long as it could be assured of a reliable source of fluorspar. For the St Lawrence Corporation, it was a different story. It had to compete on the open market to sell its product. After World War II, the St Lawrence Corporation was never able to be competitive. During World War II and the Korean War, financial support from the US government and purchase of the mine's fluorspar postponed its inevitable closure until 1957.

The second challenge was the high content of silica in the ore. After World War II technological advances in the smelting of steel and aluminum required increasingly smaller percentages of silica. This was what drove Alcan to close its St Lawrence subsidiary in 1978 and to switch to Mexican fluorspar, which had a low silica content and was also cheaper.

The unforeseen epidemic of lung cancer was the most noteworthy and widely known event in the history of the mines. It was the first of its kind reported in non-uranium mines. Several years passed before the cause was detected – radon and radon progeny from the underlying bedrock seeping into the mines in the groundwater. Augmented ventilation effectively reduced radiation to levels that the experts assured management were safe. At the time there were no standards for limits of exposure to alpha radiation. The Newfoundland authorities had to set their own limits, and when ACGIH permissible exposure limits were promulgated, they followed these. Over the years, the mines were thought to be safe. Nevertheless, in the 1990s, it was shown that risk for developing alpha radiation lung cancer is inversely proportional to the rate of accumulation. How much this adds to the risk of the St Lawrence miners for developing lung cancer has yet to be settled.

The radiation disaster required that the victims and their dependents be compensated. The Workmen's Compensation Act, in response to pressures from the public and Alcan, was amended to add cancer of the lung due to radiation as a compensable disease. The scale of benefits of the Newfoundland Workmen's Compensation Commission was comparable to those of the other provinces throughout the lifetime of the mines. The victims received full disability pensions. A special fund, financed jointly by the provincial government and Alcan was set up to provide supplementary pensions to the victims and their dependents. If the Royal Commission could not publicly endorse the level of payments of workers' compensation, it was not restrained from saying when compensation was inadequate. After conducting its own survey, the Royal Commission concluded that the pensions of widows were just sufficient to satisfy subsistence needs. Therefore, the widows had to be subsidized through the special fund. Some of them also took on jobs.

Like in so many industries, relations between the St Lawrence miners and Workmen's Compensation did not always run smoothly. There were two reasons for this state of affairs. Workmen's Compensation Boards demanded proof of a claim before accepting it. Workers were unable to understand this requirement, believing they should be given the benefit of the doubt. Partial disability pensions for non-lethal conditions were another source of resentment. Recipients were expected to look for jobs within the limits of their capabilities or, failing this, incur the stigma of applying for social assistance. Often they were unable to find or hold suitable jobs.

The epidemic of lung cancer in the St Lawrence fluorspar mines was an unforeseen event. The Royal Commission into the radiation disaster found no one to blame. The expertise required to tackle the calamity was assembled and put to work. Opinions differ on how well the epidemic was managed.

The last company to exploit the fluorspar deposits at St Lawrence, the St Lawrence Fluorspar Company Ltd was enticed to do so by government subsidies. It went bankrupt after a few years. Its short history demonstrates some of the drawbacks to government subsidized industrial projects – unwarranted optimism, weak trade unions, and cutting corners in occupational health surveillance.

Two questions remain. In the long term, is there a future for fluorspar mining at St Lawrence? The answer must surely be only if it can be competitive. The mines have recently been reactivated. Will the underground workers be at risk for developing cancer of the lung? The St Lawrence workers will not be any more at risk for developing lung cancer than those in any other mine where levels of alpha radiation have been deemed safe according to international standards. Adequate ventilation appears to be the key to providing a safe environment.

Appendices

APPENDIX A

Fluorspar: Geology, Distribution, and Industrial Uses

Fluorspar or fluorite, a mineral occurring as veins in igneous rocks, is widely distributed throughout the world. In Canada, the major deposit is at St Lawrence, which at one time produced 90 per cent of Canada's output. Smaller deposits have been worked intermittently in Quebec, Ontario, and Nova Scotia, most frequently at Madoc, Ontario the largest deposit in Ontario. Fluorspar concentrates contain substances that must be reduced because they either interfere with the industrial processing of fluorspar or pose health hazards. Silica is the prime example of the first, while arsenic, phosphorus, lead, and carbonates are the most common hazardous examples of the second. The percentages of these substances in the concentrates vary depending on where they are mined. St Lawrence ores are high in silica but low in arsenic. Mexican fluorspar is the reverse. In smelting, the focus is on silica levels, to reduce them to 1 per cent or less. In the case of industrial processes that use fluorspar in the manufacture of utensils designed for human use, such as Teflon utensils, safeguarding human health is paramount. Consequently, levels of harmful contaminants such as arsenic must be reduced to a minimum. Nowadays, desirable levels of silica and safe levels of harmful substances in fluorspar industrial processes are achieved by a mix of technical processes and blending concentrates with high levels of silica or harmful compounds with concentrates with low levels of these substances.

The chemical formula for fluorspar is CaF_2. Fluorspar for commercial purposes is categorized into three grades depending on the percentages of (CaF_2) and impurities, most importantly silica (SiO_2).

1 Acid: A minimum of 97% CaF_2 and a maximum of 1% SiO_2 used to produce hydrofluoric acid, which has many industrial uses, such as a flux in the electrolytic production of aluminum but also in aerosols, aviation fluid, refrigeration coolants, and wood preservatives.
2 Ceramic: 94–97% CaF_2 and a maximum of 2¼% SiO_2 used in the manufacture of abrasives, ceramics, and glass.
3 Metallurgical: 80–85% CaF_2 and a maximum of 5% SiO_2 expressed as effector units calculated by subtracting $2^1/_2$ times the SiO_2 percentage from the CaF_2 percentage, used as a flux in the manufacture of steel. Important requirements are to keep the SiO_2 content low and the size of the CaF_2 particles. Milling produces CaF_2 powder or "fines," which, in the open hearth smelting of steel, are lost in the smoke. Therefore, in steel production the fines content has to be 10% or less of the total CaF_2 content. The ideal particle size is $1^1/_2$ inches and must be in the form of gravel or a lump. Before World War II, 70–72% effector units of CaF_2 were acceptable to steel manufacturers. But after the war, at least 80% effector units were demanded. The St Lawrence Corporation found it increasingly difficult to meet the progressively more stringent requirements of steel manufacturers. In fact, according to the 1958 Canada Tariff Board report, the St Lawrence fluorspar deposits are more suitable for conversion into acid grade because of the high percentage of "fines" generated in producing metallurgical grade.

The technical information in this appendix is compiled from Aylward, *Report of the Royal Commission Respecting Radiation* 1:24 and various sources cited throughout this book.

APPENDIX B

St Lawrence Corporation Production Statistics

Year	Tonnage* Mined †	Acid** Grade 97%+	Metallurgical** Grade 85%+	Cyanamid** Grade 93%	Sub–metallurgical** Grade 70%+	Total Tonnage** (Short Ton)	Employees§**	Wages***	Value**** Of Shipments
1933			1,200	400		1,600		$ 4,506	
1934	NOT	302	1,288	1,204		2,794		14,179	$ 14,775
1935	KNOWN	0	1,500	3,000		4,500		26,686	18,462
1936		1,822	5,188	2,358		9,368		53,655	50,682
1937		2,172	6,125	1,049		9,346		59,905	78,596
1938		4,664	3,958	1,237		9,859	168	86,892	84,000
1939		3,569	3,969	4,838		12,376	258	98,493	60,235
1940		6,699	9,502	–		16,201	228	124,413	62,375
1941		4,448	8,318	–		12,766	224	139,802	
1942	17,676	4,969	10,856	–		15,825	209	239,489	
1943	23,084	–	14,479	–		14,479	220	273,620	
1944	18,984	4,300	19,562	–		23,862	217	304,372	
1945	22,873	5,180	15,737	–		20,917	156	271,702	
1946	20,272	2,586	8,487	1,730		12,803	122	165,840	
1947	17,315	7,384	6,996	2,173		16,553	140	147,326	
1948	23,356	13,830	10,185	–		24,015	212	274,511	
1949	37,317	15,452	9,870	–		25,322	163	248,020	

Year	*Tonnage** *Mined †*	*Acid*** *Grade* *97%+*	*Metallurgical*** *Grade* *85%+*	*Cyanamid*** *Grade* *93%*	*Sub–metallurgical*** *Grade* *70%+*	*Total Tonnage*** *(Short Ton)*	*Employees§***	*Wages****	*Value***** *Of Shipments*
1950	15,306	11,901	4,341	1,445		17,687	204	386,196	
1951	43,363	14,226	12,973	–		27,199	267	585,180	
1952	47,966	14,639	11,417	–		26,056	182	574,734	
1953	49,147	19,053	6,413	–	2,725	28,191	215	646,986	
1954	74,815	2,374	–	–	56,525	58,899	248	880,872	
1955	100,424	–	–	–	58,309	58,309	243	839,742	
1956	100,052	–	–	–	72,717	72,171	180	298,753	
1957	45,000	–	–	–	18,990	18,990		273,602	
1958	3,000								
1959	4,500								
1960	7,500								
1961	4,500								

* Report of RC, Sect. 1, Ch 1, Courtesy Department of Natural Resources, Newfoundland and Labrador.

** Donald A. Poynter, submission.

*** W.E. Seibert, brief. He gives lower figures for total tonnage than Poynter.

**** R.J. Wardle, "Pre Confederation Mineral Production by Quantity and Value, Appendix 4C," Courtesy Department of Natural Resources, Newfoundland and Labrador.

§ Includes casual labour.

† Type of ton not specified.

APPENDIX C

Newfluor Production Statistics

Year	Tonnage*§ Mined	Year	Tons**§ Shipped	Grade** Shipped CaF_2	Cost Fob St Lawrence** As Shipped Wet	Cost Fob St Lawrence** Converted To 70% Dry	Est. Cost For** Conversion To Acid Grade At Arvida [3]
1941	12,450	1965	111,873	68.9	$ 17.25	$ 18.23[2]	$ 49
1942	17,413	1966	84,266	68.6	25.35	26.71	66
1943	57,232	1967	73,384	68.2	26.69	28.70	70
1944	33,658	1968	97,125	63.7	26.43	30.78	70
1945	22,369	1969	101,420	71.6	26.42	26.84	63
1946	–	1970	157,976	69.0		32.27	72
1947	10,422	1971	86,640	66.8	32.54	35.03	77
1948	54,428	1972	157,808	63.9	36.08	41.30[2]	87
1949	34,398	1973	151,169	61.1	35.67	42.83[2]	89
1950	39,869	1974[1]	177,000	62.0	42.15	49.73[2]	101
1951	58,197	1975	182,000	62.0	46.01	54.28[2]	108
1952	87,689	1976	177,000	62.0	48.74	57.50[2]	113
1953	104,126	1977	177,000	62.0	51.63	60.91[2]	119
1954	94,997	1978	177,000	62.0	54.86	64.72[2]	125
1955	123,055	1979	177,000	62.0	58.26	68.73[2]	132
1956	118,154	1980	177,000	62.0	61.87	72.99[2]	138
1957	71,167						

Year	Tonnage*§ Mined	Year	Tons**§ Shipped	Grade** Shipped CaF_2	Cost Fob St Lawrence** As Shipped Wet	Cost Fob St Lawrence** Converted To 70% Dry	Est. Cost For** Conversion To Acid Grade At Arvida [3]
1958	77,140						
1959	66,583						
1960	83,885						
1961	109,560						
1962	138,420						
1963	134,436						
1964	144,424						
1965	148,643						
1966	124,256						
1967	128,623						
1968	150,381						

* Selected from *Report of Royal Commission Respecting Radiation,* Sect. 1 Ch. 1, Courtesy Department of Natural Resources, Newfoundland and Labrador.

** Newfluor Production Costs, 1965–80, Courtesy of Rio Tinto Alcan.

1 Figures 1974 and after [projections] from Newfluor's revised seven-year plan (1-V-74).

2 Approx. based on estimated moisture content (4.5%).

3 Assumes $4 boat and rail freight St Lawrence to Arvida; 85% flotation recovery; $5.50/ton concs. for reagents, steam, power, air, grinding balls, raw spar trucking, and other handling; $4.50/ton concs. for opr. labour, materials and R and M; $3.00/ton concs. for insurance, depreciation, and administration.

Hence, estimated cost of acid grade at Arvida

$$= (\text{Cost FOB St Lawrence} + \$4 \text{ freight}) \times \frac{97}{70 \times 0.85} + \$13.00$$

$$= (\text{Cost FOB St Lawrence} + \$4) \times 1.63 + \$13.00$$

§ Type of ton not specified.

APPENDIX D

Fluorspar Mining at St Lawrence (Total Shipments, Employment, and Value of Shipments) 1941–1978, 1987–1990*

Year	Shipment In Tonnes (Metric)	Employees	Salaries	Value of Shipment
ST LAWRENCE CORPORATION AND NEWFLUOR				
1941	14,150			$ 122,250
1942	35,247			484,876
1943	100,748		N/A	1,855,779
1944	48,568			1,089,771
1945	45,968			1,134,215
1946	20,598			202,720
1947	26,156			297,134
1948	80,906			1,103,905
1949	52,687			1,405,033
1950	20,435			1,290,361
1951	61,621			1,966,477
1952	73,739			2,484,943
1953	79,554			2,631,698
1954	107,107	423	$1,512,169	2,975,896
1955	115,561	497	1,564,553	2,678,641
1956	126,825	470	1,427,861	3,395,061
1957[§]	59,429	321	1,089,027	1,662,602

Year | Shipment In Tonnes (Metric) | Employees | Salaries | Value of Shipment

NEWFLUOR				
1958	47,351	168	$630,051	$1,483,368
1959	48,961	152	569,762	1,749,903
1960	66,011	258	941,611	1,820,769
1961	72,430	272	1,074,478	1,951,800
1962	68,039	218	880,030	1,870,184
1963	77,111	206	899,923	1,976,006
1964	87,090	214	964,660	2,254,060
1965	87,090	226	1,077,244	2,677,443
1966	87,090	252	1,215,553	1,890,768
1967	87,090	97	610,323	2,097,391
1968	87,090	233	1,270,134	2,602,230
1969	87,090	278	1,666,058	3,036,931
1970	87,090	327	1,950,665	4,595,522
1971	87,090	198	–	2,819,091
1972	87,090	368	2,745,385	5,432,151
1973	137,138	376	–	4,620,382
1974	172,683	385	3,592,522	7,119,090
1975	–	175	1,857,897	–
1976	39,811	188	2,003,761	2,934,995
1977	129,164	128	1,963,968	8,685,119
1978				6,685,119
ST LAWRENCE FLUORSPAR COMPANY LTD				
1987	10,740	56	1,539,420	1,395,003
1988	49,147	103	–	6,258,000
1989	51,624	138	3,256,491	6,422,000
1990	16,800	100	–	2,300,000

* Selected from "1933 to present – Fluorspar Mining in St Lawrence NL (shipments, employment and value of shipments)." Courtesy Department of Natural Resources, Newfoundland and Labrador.

§ The St Lawrence Corporation continued to mine small amounts of fluorspar intermittently until 1961, requiring a few workers from time to time.

APPENDIX E

Processing of Claims By Workmen's Compensation Board (WCB) in 1967

The following outlines the procedure that workers were required to follow to file a claim with the WCB. The information is paraphrased from the *Report of Royal Commission Respecting Radiation*, Sect. 1, Ch. 18, 242.

1 Employee registered a claim with the WCB.
2 Attending physician requested to send medical history and x-rays to WCB.
3 Employee's work history requested from employer.
4 These submissions evaluated by WCB:
   a) If sufficient to establish diagnosis, claim accepted, and compensation benefits awarded.
   b) If not sufficient to establish diagnosis, employee brought to St John's for examination and testing by WCB physicians. Claim could be:
      i) Accepted and compensation benefits awarded.
      ii) Not accepted. Employee had the option of appealing to the Committee of Medical Referees for final decision.

APPENDIX F

Scale of Compensation at Various Years, Workplace Health, Safety and Compensation Commission (WHSCC) 1951–2005

	Total Disability Pensions		*Survivor's Monthly Benefits*	
Year	*Percentage Rate of Salary (%)*	*Maximum Annual Salary for Computation Purposes*	*Widow*	*Child*
1951	66-2/3*	$ 3,000	$ 50.00	$ 10.00
1956	75*	3,000	60.00	20.00
1961	75*	4,000	75.00	25.00
1967	75*	5,000	100.00	35.00***
1971	75*	7,000	120.00	40.00
1976 (Jan.)	75*	10,500	250.00	60.00
1976 (Jul.)	75*	12,000	250.00	60.00
1981	75*	19,000	370.00	85.00
1984	90**	45,500	482.16	117.60
1989	90**	45,500	571.47	139.38
1993	80**	45,500	621.20	151.53
1998	80**	45,500	641.24	156.41
2005	80**	46,275	742.21	181.04

* Percentage of Gross Salary

** Percentage of Net Salary

*** 21 Years or Less

Courtesy WHSCC and Department of National Resources, Newfoundland and Labrador

Glossary of Technical Terms

ALUMINUM See Appendix A.
AMBIENT Surrounding air.
ASBESTOSIS See Pneumoconiosis(es).
BENEFICIATION Processing an ore for industrial use.
BRONCHOSCOPY The visualization of the bronchi using an instrument (bronchoscope).
CAPTIVE MINE A mine owned by an industry that determines the amount of its product.
CARBIDE (ACETYLENE) LAMP Burns acetylene gas produced by reacting calcium carbide with water.
CARCINOGENISIS Inducing cancer.
CLOSED SHOP Membership in a trade union, a condition of hiring.
COHORT An epidemiological term referring to a group of individuals having a statistical factor in common.
CONDITIONAL CLOSED SHOP Membership in a trade union, obligatory within a specified time after hiring.
COST BENEFIT ANALYSIS In occupational health refers to the cost for achieving a certain level of a hazard and the incidence of the disease caused by the hazard at that level of exposure. Sequential reductions in a hazard eventually encounter the rule of diminishing returns, i.e., the costs of further reductions of the hazard rising exponentially but diminishing numbers of exposed subjects developing the disease in question. At some point, a decision has to be made to desist from all further attempts to reduce the hazard level.
CROWN COLONY A UK colony with no legislative or administrative autonomy.

DOLE Government welfare payments during the depression in Newfoundland.

DOSE RESPONSE Risk for developing an adverse condition from exposure to a hazard directly related to the level of the hazard.

EDISON SAFETY LAMP Consists of a lightweight storage battery carried on the back connected with a light (tungsten lamp) fastened to a miner's hat. A parabolic reflector and heavy lens distributes the light in front of a miner.

FILTER CAKE The compacted solid or semi-solid material separated from a liquid after pressure filtration.

FINES Aluminum particles.

FLUORSPAR (FLUORITE) See Appendix A.

FREE ON BOARD (FOB) The sellers' costs for processing goods prior to delivery to purchaser.

FROTH FLOTATION A process whereby crushed ore is violently stirred in a mixture of surfactants and wetting agents. The fluorspar rises to the surface in a froth, leaving the gangue at the bottom.

GANGUE Worthless rock in which valuable minerals are found.

GEIGER COUNTER An instrument that measures the intensity of radiation such as particles from radioactive material.

HALF-LIFE Time to lose half of radioactivity.

HEAVY MEDIA SEPARATION (SINK AND FLOAT) A process whereby coarse ore is fed into a container filled with a medium whose specific gravity is just above that of fluorspar. The fluorspar sinks and impurities float. Replaced froth flotation in the 1940s.

HEMOPOETIC Blood cell system.

HISTOLOGY The structure of tissues and cells.

IGNEOUS Rocks brought about by intense heat.

IONS Atoms that have acquired an electric charge by gaining or losing one or more electrons.

IONIZATION Converted into ions.

JACKHAMMER Portable percussive drill operated by compressed air.

LATENCY The time lapse from earliest exposure to a hazard to the first manifestation of the disease it causes.

LINEAR RISK All levels of exposure to a hazard carry a risk.

MILLER A surface worker in a mine mill processing the ore.

MODEL Technically refers to a hypothesis constructed to explain a subject's physiological response to a hazardous agent.

MONITOR To check for compliance with an industrial regulation or for the appearance of an industrial disease.

MUCKING Loading.
MUCKING MACHINE Shovels or scrapes and loads.
ONCOLOGIST A specialist in cancer.
OUTPORT Newfoundland term for a fishing village.
PNEUMOCONIOSIS(ES) A group of chronic lung diseases caused by inhaling an industrial dust, each disease designated by the dust inhaled e.g., silicosis (silica), asbestosis (asbestos), etc. The earliest manifestation is in the chest x-ray with the appearance of multiple small nodules (round in silicosis and irregular in asbestosis) which increase in size.
QUARTZ Crystalline silica.
RADIATION Process of emission of energy or particles. Form depends on the type of emitted energy/matter.
RADON AND RADON PROGENY See uranium.
RADIOACTIVITY The spontaneous disintegration of the nucleus of an atom emitting alpha (short penetration), beta (intermediate penetration), and gamma (greatest penetration) rays.
RELATIVE RATE Expressed as a percentage.
RESPONSIBLE GOVERNMENT Self-government of a British colony in most internal matters.
ROCK BURST "Uncontrolled disruption of a highly stressed rock associated with a violent release of energy" (*Report of RC, Respecting Radiation* 2:295–310).
SCINTILLATION COUNTER An instrument for measuring the level of one of the three types of radiation (alpha, beta, or gamma).
SCHOONER A two-masted sailing vessel.
SILICA (SILICON) A compound noted for its hardness, found in hard rocks such as granite. Its chemical name is silicon dioxide (SiO_2).
SILICOSIS See Pneumoconiosis(es).
SUBCLINICAL The period in a disease before symptoms, signs, or laboratory changes become manifest.
SYNERGISM The interaction of two factors producing an effect greater than the sum of each separately.
THRESHOLD The level of a hazard below which prolonged exposure causes no adverse health effects.
TON/TONNE Short: 2,000 pounds; Long: 2,240 pounds; Metric tonne: 1,000 kilograms
URANIUM A radioactive element occurring in small amounts in many rocks, including the granite at St Lawrence. It disintegrates or

decays successively into a number of products, one being radium, which in turn decays to radon (half-life 3.825 days), and which in turn decays to radon progeny. The latter two are volatile gases. Radon progeny emit alpha particles, which, depending on the type, have a half-life of a few seconds to 26.8 minutes. Alpha radiation has little penetration and is blocked by a sheet of paper.

WASHING, SCREENING, AND PICKING Primitive process for separating fluorspar by steps listed above.

WORKING LEVEL (WL) A measurement of alpha radiation at a workplace. 1 WL (referred to in this book as the Holaday interim WL) is "any combination of radon progeny in one liter of air that results in the ultimate emission of 1.3×10^5 Mev of potential alpha energy" (RC, 1:53–66).

WORKING LEVEL MONTH (WLM) A measurement of cumulative personal exposure to alpha radiation over a period of time. Estimated by averaging locational alpha measurements (WLS) to which a worker is exposed over a period of four and one-third weeks, each week consisting of forty hours exposure multiplied by the number of months of exposure.

Notes

PREFACE

1 Sigerist, introduction to *The History of Miners' Diseases* by George Rosen, ix.
2 Aylward, *Report of Royal Commission Respecting Radiation Compensation and Safety at the Fluorspar Mines St Lawrence Newfoundland*, 1:33–42, 43–52.
3 Ibid., 53–66.
4 Rennie, *The Dirt*.
5 Hillgartner, "The Political Language of Risk," 25–6.
6 Sagoff, "Sense and Sentiment in Occupational Safety and Health," 179–97.
7 Galbraith, *New Industrial State*, 292–306.
8 Gersuny, *Work Hazards and Industrial Conflict*, 1–19.
9 Galbraith, *New Industrial State*, 292–306.
10 Leyton, *Dying Hard*.
11 Rennie, *The Dirt*.

CHAPTER ONE

1 For the geology, distribution, and industrial uses of fluorspar, see Appendix A.
2 Howse et al., *Fluorspar Deposits in the St Lawrence Area*.
3 Anonymous, personal communication with author.
4 Martin, *Once Upon a Mine*, 52–65, 74–82.
5 There was also to be another fluorspar mining company, Newfluor, which will be discussed later in this volume.

6 "St Lawrence to Point May," *Decks Awash*, (November–December 1984), 5–12.
7 Martin, *Once Upon a Mine*, 66–72.
8 Ibid.; Slaney, *More Incredible than Fiction*.
9 Martin, *Once Upon a Mine*, 66–72.
10 *RC Respecting Radiation,* Newfoundland Fluorspar Ltd Submission to RC, July 1967, Box III, 131, 1–14.
11 Martin, *Once Upon a Mine*, 66–72.
12 Slaney, *More Incredible than Fiction*.
13 Donald A. Poynter, "Submission Supporting a Tariff on All Grades of Fluorspar Entering Canada," April 1958. Joseph R. Smallwood Papers, 3:20:087, Memorial University of Newfoundland, St John's, NL.
14 Slaney, *More Incredible than Fiction*; Snelgrove, *Mines and Mineral Resources of Newfoundland*.
15 "St Lawrence to Point May," *Decks Awash*, (November–December 1984), 5–12; Slaney, *More Incredible than Fiction*.
16 Slaney, *More Incredible than Fiction;* Edwards and Buehler, *Notes Toward a History of St Lawrence*, 58–69.
17 Slaney, *More Incredible than Fiction*.
18 *RC Respecting Radiation,* Newfoundland Fluorspar Ltd Submission to RC, July 1967, Box III, 131, 15–18.
19 "Submission Supporting a Tariff," Joseph R. Smallwood Papers, 3:20:087.
20 Thompson, "Time, Work-Disciplining and Industrial Capitalism," 39–77; Leyton, *Dying Hard*, 20–32, 88–99.
21 *Commission of Government, Secretary of the Commission*. Memorandum submitted by the Commissioner of Natural Resources for consideration by Commission of Government, 13 November 1935. S2-1-11, NR 31-35, 36, File 1.
22 Aylward, *Report of Royal Commission Respecting Radiation*, 1:17–32.
23 Adams, "The Forgotten Miners," 21–46.
24 Memorandum submitted by the Commissioner of Natural Resources for consideration by Commission of Government, 13 November 1935. Commission of Government, Secretary of the Commission Fonds, S2-1-11, NR 31-35, 36, File 1. Later Dr Smith was to work for Newfluor as manager.
25 "St Lawrence to Point May," *Decks Awash*, (November–December 1984), 5–12.
26 Slaney, *More Incredible than Fiction*.
27 Fraser, *Report of the St Lawrence Trade Dispute Board*.
28 Liddell, *Industrial Survey of Newfoundland*.
29 Martin, *Once Upon a Mine*, 52–65, 74–82.

30 Higgins, *Report of the Mining Industry of Newfoundland.*
31 Howse, "Our Story."
32 Corlett, *Inspection of Newfoundland Mines for Safety of Workmen and Operations*, 1949.
33 Memorandum submitted by Commissioner of Natural Resources for consideration by Commission of Government with attachment Report on Investigation of conditions at St Lawrence by C.K. Howse, 1 June 1937. Commission of Government, Secretary of the Commission Fonds, S2-1-11, NR 44-137, File 3 and S5-4-1, File 5.
34 Howse, "Our Story." Howse was to go on to a distinguished career as government geologist, deputy minister of mines (1949–55), and then administrative positions in BRINCO and IOCC.
35 Ibid.
36 Memorandum submitted by the Commissioner of Natural Resources for consideration by Commission of Government, 13 November 1935, Commission of Government, Secretary of the Commission Fonds, S2-1-11, NR 31-35, 36, File 1.
37 If this was based on total tonnage, I calculate from Appendix B, it would have given him an income of $27,000 in 1935.
38 Memorandum submitted by the Commissioner of Natural Resources for consideration by Commission of Government, 13 November 1935, Commission of Government, Secretary of the Commission Fonds, S2-1-11, NR 31-35, 36, File 1.
39 Memorandum submitted by Commissioner of Natural Resources for consideration by Commission of Government with attachment Report on Investigation of conditions at St Lawrence by C.K. Howse, 1 June 1937, Commission of Government, Secretary of the Commission Fonds, S2-1-11, NR 44-137, File 3 and GN-38, S5-4-1, File 5.
40 L.J. Saint to Mr Stevenson, Credit Department, Royal Stores, 9 June 1937, Commission of Government, Secretary of the Commission Fonds, S5-4-1, File 5.
41 Martin, *Once Upon a Mine*, 66–72.
42 Ibid.; Aylward, *Report of Royal Commission Respecting Radiation*, 1:17–32.
43 *RC Respecting Radiation*, Newfoundland Fluorspar Ltd. Submission to RC, July 1967, Box III, 131, 15–18.
44 Slaney, *More Incredible than Fiction.*
45 Martin, *Once Upon a Mine*, 66–72.
46 Howse et al., *Fluorspar Deposits in the St Lawrence Area.*
47 Slaney, *More Incredible than Fiction.*
48 Martin, *Once Upon a Mine*, 66–72.

49 Slaney, *More Incredible than Fiction.*
50 Corlett, *Inspection of Newfoundland Mines for Safety of Workmen and Operations*, 1949.
51 Martin, *Once Upon a Mine*, 66–72.

CHAPTER TWO

1 Memo, Commissioner of Public Utilities, 23 March 1944, Commission of Government, Secretary of the Commission Fonds, S5-4-1, File 6; Howse, "Our Story"; "Submission Supporting a Tariff," Joseph R. Smallwood Papers, 3:20:087.
2 "Submission Supporting a Tariff," Joseph R. Smallwood Papers, 3:20:087.
3 Martin, *Once Upon a Mine*, 66–72.
4 Howse, "Our Story."
5 Hattenhauer, *A Brief Labour History of Newfoundland.*
6 Ibid.
7 Kealey, *The History and Structure of the Newfoundland Labour Movement.*
8 H. M. Mosdell to Department of Justice, 13 January 1940, Commission of Government, Secretary of the Commission Fonds, S6-1-2, File 12.
9 Fraser, *Report of the St Lawrence Trade Dispute Board.*
10 Ibid.
11 Hattenhauer, *A Brief Labour History of Newfoundland.*
12 Ibid.
13 Liddell, *Industrial Survey of Newfoundland.*
14 Gillespie, *A Class Act.*
15 Liddell, *Industrial Survey of Newfoundland.*
16 Secretary of State Dominion Affairs to Government of Newfoundland, 11 February 1940, Commission of Government, Secretary of the Commission Fonds, S5-6-1, File 9.
17 Thompson, *The Making of the English Working Class*, 281.
18 Liddell, *Industrial Survey of Newfoundland.*
19 Martin, *Once Upon a Mine*, 66–72; Weir, *The Miners of Wabana*, 22–30.
20 Weir, *The Miners of Wabana*, 22–30.
21 Edwards and Buehler, *Notes Toward a History of St Lawrence*, 58–69; Aylward, *Report of Royal Commission Respecting Radiation*, 1:17–32.
22 Report by Magistrate N. Short, J.P. to W.W. Woods, Commissioner of Public Utilities, 30 April 1940, Commission of Government, Secretary of the Commission Fonds, S5-4-1, File 5.
23 Fraser, *Report of the St Lawrence Trade Dispute Board.*

24 Report by Magistrate N. Short, J.P. to W.W. Woods, Commissioner of Public Utilities, 30 April 1940, Commission of Government, Secretary of the Commission Fonds, S5-4-1, File 5.
25 Fraser, *Report of the St Lawrence Trade Dispute Board.*
26 Ibid.
27 Ibid.
28 Ibid.
29 Ibid.; Rennie, "The Historical Origins of an Industrial Disaster," 107–42.
30 Fraser, *Report of the St Lawrence Trade Dispute Board.*
31 Ibid.
32 Rennie, "The Historical Origins of an Industrial Disaster," 107–42; Sir Walter W. Woods to St Lawrence Workers Protective Union, 8 September 1941, Commission of Government, Secretary of the Commission Fonds, S5-4-1, File 5.
33 *Commission of Government, Secretary of the Commission.* Sir Walter W. Woods to St Lawrence Workers Protective Union, 8 September 1941, S5-4-1, File 5.
34 Fraser, *Report of the St Lawrence Trade Dispute Board.*
35 Ibid.
36 Walter E. Seibert to Sir Walter W. Woods, 22 July 1941, Commission of Government, Secretary of the Commission Fonds, S5-6-1, File 5.
37 Donald A. Poynter to Sir Walter W. Woods, 28 October 1941, Commission of Government, Secretary of the Commission Fonds, S5-6-1, File 5.
38 Fraser, *Report of the St Lawrence Trade Dispute Board.*
39 P.J. Lewis to Sir Walter W. Woods, 11 November 1941. Commission of Government, Secretary of the Commission Fonds, S5-6-1, File 5.
40 Chas. E. Hunt to Sir Walter W. Woods, 6 November 1941. Commission of Government, Secretary of the Commission Fonds, S5-6-1, File 5.
41 C.J. Burchell to Sir Walter W. Woods, 7 November 1941. Commission of Government, Secretary of the Commission Fonds, S5-6-1, File 5; Gillis, telegram, n.d., no recipient named. Commission of Government, Secretary of the Commission Fonds, S5-4-1, File 5.
42 Sir Walter W. Woods to P.J. Lewis, 10 November 1941. Commission of Government, Secretary of the Commission Fonds, S5-6-1, File 5.
43 Fraser, *Report of the St Lawrence Trade Dispute Board.*
44 Ibid.
45 Ibid.
46 Alec T. Hickman, personal communication.
47 Rennie, "The Historical Origins of an Industrial Disaster," 107–42.

48 A.J. Walsh, *Report of the Labour Relations Office from 1 June 1942–8 February 1944*, Commission of Government, Secretary of the Commission Fonds, S5-6-1, File 5.
49 Fraser, *Report of the St Lawrence Trade Dispute Board.*
50 H.M. Mosdell to Department of Justice, 13 January 1940. Commission of Government, Secretary of the Commission Fonds, S6-1-2, File 12; Slaney, *More Incredible than Fiction.*
51 Report by Magistrate N. Short, J.P. to W.W. Woods, Commissioner of Public Utilities, 30 April 1940. Commission of Government, Secretary of the Commission Fonds, S5-4-1, File 5.
52 Fraser, *Report of the St Lawrence Trade Dispute Board*; Rennie, "The Historical Origins of an Industrial Disaster," 107–42.
53 Fraser, *Report of the St Lawrence Trade Dispute Board.*
54 Ibid.; Aylward, *Report of Royal Commission Respecting Radiation*, 1:161–6.
55 Martin, *Leonard Albert Miller Public Servant*, 49–51.
56 H. M. Mosdell to Department of Justice, 13 January 1940. Commission of Government, Secretary of the Commission Fonds, S6-1-2, File 12.
57 Fraser, *Report of the St Lawrence Trade Dispute Board.*
58 Muir et al., "Silica Exposure and Silicosis," 5–11.
59 Fraser, *Report of the St Lawrence Trade Dispute Board.*
60 At the time of the trade dispute board, closed shop referred to the fact that workers elected a trade union to negotiate with management on its behalf. Membership in the union was not always obligatory. In unconditional closed shop, membership in the union was obligatory for all new workers. In conditional closed shop, certain individuals and categories of workers were exempt from union membership.
61 Fraser, *Report of the St Lawrence Trade Dispute Board.*
62 Nowadays expert opinion would side with the workers.
63 Gillespie, *A Class Act.*
64 Sir H. Walwyn to Clement Attlee, 5 August 1942 and 6 November 1942. Commission of Government, Secretary of the Commission Fonds, S5-4-1.
65 A.J. Walsh, *Report of the Labour Relations Office from 1 June 1942–8 February 1944*. Commission of Government, Secretary of the Commission Fonds, S5-6-1, File 5.
66 Slaney, *More Incredible than Fiction.*
67 "St Lawrence to Point May," *Decks Awash*, (November–December 1984), 45–8.
68 Reports stated that the facilities met required standards. Slaney notes in his book that this was not the case. A.J. Walsh, *Report of the Labour*

Relations Office from 1 June 1942–8 February 1944. Commission of Government, Secretary of the Commission Fonds, S5-6-1, File 5.

69 A.J. Walsh, *Report of the Labour Relations Office from 1 June 1942–8 February 1944*. Commission of Government, Secretary of the Commission Fonds, S5-6-1, File 5.

70 Ibid.

71 Memo, Commissioner of Public Utilities, 23 March 1944, Commission of Government, Secretary of the Commission Fonds, S5-6-1, File 6.

72 Howse, "Our Story."

73 Higgins, *Report of the Mining Industry of Newfoundland.*

74 Peat, Marwick, Mitchell and Co., *St Lawrence Corporation of Newfoundland Limited.*

75 St Lawrence Fluorspar Inc is not to be confused with the much later St Lawrence Fluorspar Company Ltd, the third active company to mine St Lawrence fluorspar. See chapter 8. Ibid.

76 Walter E. Seibert to J. Cusick, 23 December 1949. Joseph R. Smallwood Papers, 3:20:087.

77 Walter E. Seibert to Joseph R. Smallwood, 7 November 1949. Joseph R. Smallwood Papers, 3:20:087; Wood of Richardson Wood and Company, Memo, 30 March 1950. Joseph R.Smallwood Papers, 3:20:087.

78 Walter E. Seibert to J. Cusick, 23 December 1949. Joseph R. Smallwood Papers, 3:20:087.

79 Higgins, *Report of the Mining Industry of Newfoundland.*

80 Aylward, *Report of Royal Commission Respecting Radiation*, 1:17–32.

81 Corlett, *The Safety of Workmen in Newfoundland Mining Operations*, 1948.

82 Higgins, *Report of the Mining Industry of Newfoundland.*

83 Corlett, *Inspection of Newfoundland Mines for Safety of Workmen and Operations*, 1949.

84 Corlett, *The Safety of Workmen in Newfoundland Mining Operations*, 1948.

85 Ibid.

86 Fishback and Kantor, *A Prelude to the Welfare State*, 54–87.

87 Corlett, *The Safety of Workmen in Newfoundland Mining Operations*, 1948.

88 Corlett, *Inspection of Newfoundland Mines for Safety of Workmen and Operations*, 1949.

89 Higgins, *Report of the Mining Industry of Newfoundland*; Corlett, *Inspection of Newfoundland Mines for Safety of Workmen and Operations*, 1949.

90 Corlett, *Inspection of Newfoundland Mines for Safety of Workmen and Operations*, 1950.

CHAPTER THREE

1 Fogwill, *How Modern Workmen's Compensation Law Came to Newfoundland.*
2 Gillespie, *A Class Act.*
3 Fogwill, *How Modern Workmen's Compensation Law Came to Newfoundland.*
4 Aylward, *Report of Royal Commission Respecting Radiation*, 1: 17–32.
5 Howse, *Fluorspar Deposits in the St Lawrence Area.*
6 Aylward, *Report of Royal Commission Respecting Radiation*, 1:17–32; Hodge, *Report on an Appraisal of the Alcan Production Operations.*
7 Phonce Cooper, personal communication.
8 Aylward, *Report of Royal Commission Respecting Radiation*, 1:17–32.
9 Frank G. Barker to Joseph R. Smallwood, 30 July 1964, Joseph R. Smallwood Papers, 3:20:087.
10 Department of Natural Resources, E 1086–1178; Bassler, *Alfred Valdmanis and the Politics of Survival*, 312–13.
11 Editorial, "New Fluorspar Developments on Burin Peninsula," 23.
12 Slaney, *More Incredible than Fiction.*
13 Aylward, *Report of Royal Commission Respecting Radiation*, 1:17–32.
14 Following the TDB in 1942, St Lawrence had been able to attract a series of solo physicians, subsidized by the community and the mining companies. None stayed for more than a year until the arrival of Dr J.J. Pepper in 1949, who stayed until 1954.
15 Aylward, *Report of Royal Commission Respecting Radiation*, 1:161–6.
16 Aylward, *Report of Royal Commission Respecting Radiation*, 1:33–42.
17 Fraser, *Report of the St Lawrence Trade Dispute Board.*
18 Aylward, *Report of Royal Commission Respecting Radiation*, 1:199–208.
19 Ibid.
20 Ibid., 1:161–6.
21 Corlett, *Inspection of Newfoundland Mines for Safety of Workmen and Operations*, 1950.
22 Ibid.; Aylward, *Report of Royal Commission Respecting Radiation*, 1:185–98.
23 Aylward, *Report of Royal Commission Respecting Radiation*, 1:185–98.
24 Ibid., 1:161–6.
25 Ibid.
26 Ibid., 1:33–42.
27 Ibid., 1:161–6.

28 Not until 1960 was every miner required to hold a Certificate of Good Health, which had to be renewed annually. A chest x-ray was one of its requirements.

29 Rosner and Markowitz, *Deadly Dust,* 75–104.

30 Gravimetric sampling, a more accurate means of measuring dust levels, is now the preferred choice.

31 Rosner and Markowitz, *Deadly Dust,* 75–104.

32 Martin, *Leonard Albert Miller, Public Servant,* 157–64.

33 Alyward, *Report of Royal Commission Respecting Radiation,* 1:33–42.

34 Ibid.

35 How many of these cases were in the mill where conditions were drier, and therefore posed a greater risk for silicosis and how many in the mines, is not recorded.

36 Parsons et al., "Lung Cancer in a Fluorspar Mining Community II," 110–16.

37 Aylward, *Report of Royal Commission Respecting Radiation,* 1:199–208; 1:177–84.

38 Ibid., 1:199–208.

39 *RC Respecting Radiation,* Submission of Alcan to RC Respecting Radiation, Testimony of Frank deN. Brent, Box II, 124.

40 Evidently, the survey was to include more than dust measurements; it was also to include an epidemiological survey of the health of the miners. Alyward, *Report of Royal Commission Respecting Radiation,* 1:33–42.

41 Ibid.

42 Windish and Sanderson, *Dust Hazards in the Mines of Newfoundland.*

43 The Corporation was about to close (see Chapter 5). Consequently, Newfluor was left alone to face what was to prove a very serious situation. Frank deN. Brent, MD, *Report on a Problem in Industrial Health and Hygiene in St Lawrence, Newfoundland,* 14 February 1958, Alcan Archives Collection.

44 The hospital was a gift from the US government to the people of St Lawrence and Lawn in gratitude for their heroic measures in rescuing and caring for the surviving members of the crews of the destroyers, USS *Pollux* and USS *Truxton,* which had been wrecked at nearby Chambers Cove, 18 February 1942. Successive bills to earmark funds to build the hospital died on the floor of the House of Representatives. Eventually, after some prodding from St Lawrence and the Canadian ambassador, the promise was fulfilled in 1954. The Department of Health was not enthused at having to operate another hospital in an area that already had two cottage hospitals (Burin and Grand Bank). Aylward, *Report of Royal*

Commission Respecting Radiation, 1:161–6; St Lawrence Corporation and Newfoundland Fluorspar Limited, Library and Archives Canada, RG 39-1, vol. 38, file 35-02-0.

45 Aylward, *Report of Royal Commission Respecting Radiation*, 1:161–6.

46 Ibid., 1:99–123.

47 Doll and Hill, "The Mortality of Doctors in Relation to Their Smoking Habits," 1451–5.

48 C.J. Walsh, personal communication.

49 deVilliers and Windish, "Lung Cancer in a Fluorspar Mining Community," 94–109.

50 Frank deN. Brent, MD, "Health Problems in the St Lawrence Mine," 24 February 1967, Alcan Archives Collection.

51 Frank deN. Brent, MD, *Report on a Problem in Industrial Health and Hygiene in St Lawrence, Newfoundland*, 14 February 1958, Alcan Archives Collection.

52 Nowadays referred to as radon progeny.

53 Aylward, *Report of Royal Commission Respecting Radiation*, 1:83–97.

54 Alcan Archives Collection. Frank deN. Brent, MD to Rupert Wiseman, (no date).

55 *RC Respecting Radiation*, Submission of Alcan to RC Respecting Radiation, testimony of Frank deN. Brent, Box II, 124; *RC Respecting Radiation*, Newfoundland Fluorspar Ltd Submission to RC Respecting Radiation, brief, Box III, 131, 87–152.

56 Frank deN. Brent to Leonard Albert Miller, 18 July 1958, Alcan Archives Collection.

57 *RC Respecting Radiation*, Submission of Alcan to RC Respecting Radiation, testimony of Frank deN. Brent, Box II, 124.

58 Ibid.

59 Frank deN. Brent to Leonard Albert Miller, 18 July 1958, Alcan Archives Collection.

60 Ibid.

61 *RC Respecting Radiation*. Submission of Alcan to RC Respecting Radiation, testimonies of Rupert Wiseman and J.W. Cameron, Box II, 124, 125.

62 Ibid.

63 Alyward, *Report of Royal Commission Respecting Radiation*, 1:33–42; 1:99–123.

64 Alyward, *Report of Royal Commission Respecting Radiation*, 1:33–42.

65 *RC Respecting Radiation*, Submission of Alcan to RC Respecting Radiation, testimonies of Rupert Wiseman and J.W. Cameron, Box 2, 124, 125.

66 *RC Respecting Radiation*, Newfoundland Fluorspar Ltd, Submission to RC Respecting Radiation, brief, Box III, 131, 87–152.
67 Alyward, *Report of Royal Commission Respecting Radiation*, 1:33–42; 1:43–57; *RC Respecting Radiation*, Submission of Alcan to RC Respecting Radiation, testimonies of Rupert Wiseman and J.W. Cameron, Box II, 124, 125.
68 Alyward, *Report of Royal Commission Respecting Radiation*, 1:33–42.
69 *RC Respecting Radiation*, Submission of Alcan to RC Respecting Radiation, testimony of Frank deN. Brent, Box II, 124.
70 Aylward, *Report of Royal Commission Respecting Radiation*, 1:83–97; ACHRE, *Final Report: Advisory Committee on Human Radiation Experiments.*
71 ACHRE, *Final Report: Advisory Committee on Human Radiation Experiments.*
72 Frank deN. Brent, MD, *Report on a Problem in Industrial Health and Hygiene in St Lawrence, Newfoundland*, 14 February 1958. Alcan Archives Collection.
73 *RC Respecting Radiation,* Submission of Alcan to RC Respecting Radiation, testimony of Frank deN. Brent, Box II, 124.
74 Ibid.
75 Patterson, "NHW Occupational Health Division to Leonard Albert Miller."
76 ACHRE, *Final Report: Advisory Committee on Human Radiation Experiments.*
77 Alyward, *Report of Royal Commission Respecting Radiation*, 1:33–42.
78 *RC Respecting Radiation*, Newfoundland Fluorspar Ltd Submission to RC Respecting Radiation, brief, Box III, 131, 87–152.
79 "Government Officials Meet to Discuss Radiation Hazard in Flourspar (*sic*) Mines," *Daily News*, 2 March 1960.
80 Alyward, *Report of Royal Commission Respecting Radiation*, 1:33–42.
81 *RC Respecting Radiation*, Cross–Exam Witnesses, S.T. Payne, Box III, 127.
82 Aylward, *Report of Royal Commission Respecting Radiation*, 1:43–57.
83 deVilliers and Windish, "Lung Cancer in a Fluorspar Mining Community," 94–109.
84 Slaney, *More Incredible than Fiction.*
85 Aylward, *Report of Royal Commission Respecting Radiation*, 1:99–123.
86 Rennie, *The Dirt*, 91–107.
87 deVilliers and Windish, "Lung Cancer in a Fluorspar Mining Community," 94–109.
88 Slaney, *More Incredible than Fiction*; Parsons et al., "Lung Cancer in a Fluorspar Mining Community," 110–16; Aylward, *Report of Royal Commission Respecting Radiation*, 1:177–84; *RC Respecting Radiation*,

Submission of Alcan to RC Respecting Radiation, testimony of Frank deN. Brent, Box II, 124.

89 Aylward, *Report of Royal Commission Respecting Radiation*, 1:273–80.

CHAPTER FOUR

1 A. Turpin to Joseph R. Smallwood, telegram, 12 March 1960, Alcan Archives Collection.

2 *RC Respecting Radiation*, Newfoundland Fluorspar Ltd Submission to RC Respecting Radiation, brief, Box III, 131, 15–18.

3 Ibid.; Editorial, "St Lawrence and The Press," *Evening Telegram* (St John's), 1 December 1967.

4 *RC Respecting Radiation*, Newfoundland Fluorspar Ltd Submission to RC Respecting Radiation, brief, Box III, 131, 15–18.

5 Aylward, *Report of Royal Commission Respecting Radiation*, 1:17–32.

6 Frank G. Barker, memo, 5 April 1960, Alcan Archives Collection.

7 Frank deN. Brent to Cyril J. Walsh, MD, 9 May 1960, Alcan Archives Collection.

8 Cyril J. Walsh, MD to Frank deN. Brent, 18 May 1960, Alcan Archives Collection.

9 Ibid.

10 "Not Convinced Fluorspar Mines are Safe: Union Official Says Miners Being Used as 'Guinea Pigs,'" *Evening Telegram,* 27 February 1967.

11 *RC Respecting Radiation*, Cross-Exam Witnesses, S.T. Payne, Vice President Confederation of National Trade Unions, Box III, 127.

12 "Not Convinced Fluorspar Mines are Safe," *Evening Telegram*, 27 February 1967.

13 Adams, "The Forgotten Miners," 21–46.

14 "Not Convinced Fluorspar Mines are Safe," *Evening Telegram*, 27 February 1967.

15 Leyton, *Dying Hard*, 20–32.

16 H.A. Etheridge, memo, 14 March 1960, Alcan Archives Collection.

17 Frank G. Barker, memo, no date, Alcan Archives Collection.

18 Ibid.

19 Aylward, *Report of Royal Commission Respecting Radiation*, 1:221–72.

20 deVilliers and Windish, "Lung Cancer in a Fluorspar Mining Community," 94–109.

21 Aylward, *Report of Royal Commission Respecting Radiation*, 1:221–72.

22 Ibid.

23 Memo, 1973, Alcan Archives Collection.

24 Frank deN. Brent, Memo, 1 April 1960, Alcan Archives Collection.
25 Aylward, *Report of Royal Commission Respecting Radiation*, 1:99–123.
26 Cyril J. Walsh, MD to Frank deN. Brent, 18 May 1960, Alcan Archives Collection.
27 Aylward, *Report of Royal Commission Respecting Radiation*, 1:33–42; Dyer, *Report Number Two of the Industrial Inquiry Commission.*
28 Aylward, *Report of Royal Commission Respecting Radiation*, 1:33–42; *RC Respecting Radiation*, Cross-Exam Witnesses, Rupert Wiseman and J.W. Cameron, Box III, 127.
29 *RC Respecting Radiation*, Newfoundland Fluorspar Ltd Submission to RC Respecting Radiation, brief, Box III, 131, 169–75.
30 M.E. Gooding to Rupert Wiseman, 16 May 1968, Alcan Archives Collection.
31 Frank G. Barker to Joseph R. Smallwood, 30 July 1964, Alcan Archives Collection.
32 *RC Respecting Radiation*, Cross-Exam Witnesses, S.T. Payne, Vice President Confederation of National Trade Unions, Box III, 127.
33 Trist, *The Situation at Newfoundland Fluorspar Ltd.*
34 "Miners Get Wage Raise," *Daily News*, 13 March 1963.
35 *RC Respecting Radiation*, Newfoundland Fluorspar Ltd Submission to RC Respecting Radiation, brief, Box III, 131, 74–86; ACHRE, *Final Report: Advisory Committee on Human Radiation Experiments.*
36 ACHRE, *Final Report: Advisory Committee on Human Radiation Experiment.*
37 Ibid.; Aylward, *Report of Royal Commission Respecting Radiation*, 1:83–97.
38 ACHRE, *Final Report: Advisory Committee on Human Radiation Experiments.*
39 Ibid.
40 Ibid.
41 Aylward, *Report of Royal Commission Respecting Radiation*, 1:99–123.
42 ACHRE, *Final Report: Advisory Committee on Human Radiation Experiments.*
43 Ibid.
44 Ibid.; Aylward, *Report of Royal Commission Respecting Radiation*, 1:83–97.
45 Now it is the International Commission on Radiological Protection, ACGIH, or NIOSH that recommends a permissible exposure; the federal government endorses it, and the provincial government enforces it.
46 Rennie, *The Dirt*, 70–90.
47 Aylward, *Report of Royal Commission Respecting Radiation*, 1:53–65.

48 Ibid.
49 Edwards, *Health Effects of Ionizing Radiation*, Section B.
50 Archer, "Health Concerns in Uranium Mining and Milling," 502–5.
51 Aylward, *Report of Royal Commission Respecting Radiation*, 1:53–65.
52 "Not Convinced Fluorspar Mines are Safe," *Evening Telegram*, 27 February 1967; Frank deN. Brent to Frank G. Barker, 24 April 1960. Alcan Archives Collection.
53 Aylward, *Report of Royal Commission Respecting Radiation*, 1:53–65.
54 Ibid., 1:67–81.
55 Frank deN. Brent to Frank G. Barker, 11 July 1960, Alcan Archives Collection.
56 Gerald Drover, personal communication.
57 Aylward, *Report of Royal Commission Respecting Radiation*, 1:67–81.
58 Ibid., 1:53–65.
59 Ibid.
60 Aylward, *Report of Royal Commission Respecting Radiation*, 1:67–81.
61 Ibid.
62 *RC Respecting Radiation*, Cross-Exam Witnesses, Rupert Wiseman, Box III, 127.
63 *RC Respecting Radiation*, Cross-Exam Witnesses, S.T. Payne, Vice President Confederation of National Trade Unions, Box III, 127; *Newfoundland Herald*, 29 September 1963.
64 "Not Convinced Fluorspar Mines are Safe," *Evening Telegram*, 27 February 1967.
65 *RC Respecting Radiation*, Cross-Exam Witnesses, Letter from Walter J. Keough to P.J. Lewis, 18 March 1964, Box III, 127.
66 *RC Respecting Radiation*, Cross-Exam Witnesses, S.T. Payne, Vice President Confederation of National Trade Unions, Box III, 127; *Newfoundland Herald*, 29 September 1963.
67 "House of Assembly," *Evening Telegram*, 5 June 1964.
68 Adams, "The Forgotten Miners," 21–46.
69 Ibid.; Horwood, "Mining as a Way of Death," 4–7.
70 Leyton, "Bureaucratization of Anguish."
71 Aylward, *Report of Royal Commission Respecting Radiation*, 1:221–72.
72 Ibid.
73 Ibid.
74 Ibid.
75 Ibid.
76 Horwood, "Mining as a Way of Death," 4–7.
77 "Hickman Would Welcome Cancer Incidence Inquiry," *Evening Telegram*, 17 February 1967.

78 Aylward, *Report of Royal Commission Respecting Radiation*, 1:53–65.
79 Winter, *Report on the Workman's Compensation Act.*
80 Slaney, *More Incredible than Fiction.*
81 Winter, *Report on the Workman's Compensation Act.*
82 "House of Assembly," *Evening Telegram*, 5 June 1964.
83 "The St Lawrence Inquiry," *Evening Telegram*, 11 December 1967.
84 "Not Convinced Fluorspar Mines are Safe," *Evening Telegram*, 27 February 1967.
85 Ibid.
86 Alec T. Hickman, personal communication.

CHAPTER FIVE

1 Walter E. Seibert to Joseph R. Smallwood, 7 November 1949, Joseph R. Smallwood Papers, 3:20:087.
2 Peat, Marwick, Mitchell and Co., *St Lawrence Corporation of Newfoundland Limited.*
3 Donald A. Poynter to Joseph R. Smallwood, 9 November 1949, Joseph R. Smallwood Papers, 3:20:087.
4 Walter E. Seibert to Joseph R. Smallwood, 7 November 1949, Joseph R. Smallwood Papers, 3:20:087; Donald A. Poynter to Joseph R. Smallwood, 9 November 1949, Joseph R. Smallwood Papers, 3:20:087.
5 Walter E. Seibert to Joseph R. Smallwood, 7 November 1949, Joseph R. Smallwood Papers, 3:20:087.
6 Peat, Marwick, Mitchell and Co., *St Lawrence Corporation of Newfoundland Limited*; R. Wood of Richardson Wood and Co., "St Lawrence Corporation of Newfoundland Limited and St Lawrence Fluorspar Inc, 1950," Joseph R. Smallwood Papers, 3:20:087.
7 R. Wood of Richardson Wood and Co., "St Lawrence Corporation of Newfoundland Limited and St Lawrence Fluorspar Inc, 1950," Joseph R. Smallwood Papers, 3:20:087.
8 Peat, Marwick, Mitchell and Co., *St Lawrence Corporation of Newfoundland Limited.*
9 Ibid.
10 Joseph R. Smallwood to Walter E. Seibert, 27 October 1949, Joseph R. Smallwood Papers, 3:20:087.
11 Peat, Marwick, Mitchell and Co., *St Lawrence Corporation of Newfoundland Limited.*
12 Joseph R. Smallwood to Walter E. Seibert, 2 February 1950, Joseph R. Smallwood Papers, 3:20:087.

13 Donald A. Poynter, "Submission Supporting a Tariff," Joseph R. Smallwood Papers, 3:20:087.
14 Canada Tariff Board, *Report by the Tariff Board.*
15 This stipulation that the fluorspar be shipped to the Delaware plant for further upgrading seems to have been a strange stipulation. The Corporation had always produced acid grade fluorspar at St Lawrence. Early in World War II the US government had provided a loan to build a new froth flotation plant to produce acid grade filter cake (see chapter 2). To stop doing so seems wasteful.
16 Donald A. Poynter, "Submission Supporting a Tariff," Joseph R. Smallwood Papers, 3:20:087.
17 Walter E. Seibert to Joseph R. Smallwood, 27 December 1955, Joseph R. Smallwood Papers, 3:20:087.
18 Seibert, *Brief Supporting a Tariff Request.*
19 Donald A. Poynter, "Submission Supporting a Tariff," Joseph R. Smallwood Papers, 3:20:087.
20 Seibert, *Brief Supporting a Tariff Request.*
21 Walter E. Seibert to Joseph R. Smallwood, 27 December 1955, Joseph R. Smallwood Papers, 3:20:087.
22 Ibid.
23 Walter E. Seibert to G.H. Glass, 22 July 1957, Joseph R. Smallwood Papers, 3:20:087.
24 Walter E. Seibert to Joseph R. Smallwood, 27 December 1955, Joseph R. Smallwood Papers, 3:20:087.
25 Walter E. Seibert to G.H. Glass, 22 July 1957, Joseph R. Smallwood Papers, 3:20:087.
26 Canada Tariff Board, *Report by the Tariff Board.*
27 Donald A. Poynter, "Submission Supporting a Tariff," Joseph R. Smallwood Papers, 3:20:087; Seibert, *Brief Supporting a Tariff Request.*
28 Herbert Slaney, personal communication.
29 Canada Tariff Board, *Report by the Tariff Board.*
30 Seibert, *Brief Supporting a Tariff Request.*
31 Canada Tariff Board, *Report by the Tariff Board.*
32 Walter E. Seibert to Gordon Pushie, Director General, Department of Economic Development, Newfoundland, 16 May 1958, Joseph R. Smallwood Papers, 3:20:088.
33 Walter E. Seibert to Joseph R. Smallwood, 11 July 1957, Joseph R. Smallwood Papers, 3:20:087.
34 Joseph R. Smallwood to Walter E. Seibert, 5 February 1958, Joseph R. Smallwood Papers, 3:20:087.

35 Report on the St Lawrence Corporation by G.K. Goundrey, no date. Joseph R. Smallwood Papers, 3:20:087.
36 A. Turpin to Walter E. Seibert, 24 January 1958, Joseph R. Smallwood Papers, 3:20:088.
37 Conference on Closure of the St Lawrence Corporation, 19 February 1958, Joseph R. Smallwood Papers, 3:20:087.
38 Canada Tariff Board, *Report by the Tariff Board.*
39 Walter J. Keough to Walter E. Seibert, 24 March 1958, Joseph R. Smallwood Papers, 3:20:087.
40 Canada Tariff Board, *Report by the Tariff Board.*
41 Walter E. Seibert to Joseph R. Smallwood, 27 December 1955, Joseph R. Smallwood Papers, 3:20:087.
42 Gordon F. Pushie to Joseph R. Smallwood, 3 November 1958, Joseph R. Smallwood Papers, 3:20:087.
43 John McClurg, president, Seaforth Mineral and Ore Co., Cleveland, Ohio, to Joseph R. Smallwood, 10 September 1959, Joseph R. Smallwood Papers, 3:20:087.
44 Donald. A. Poynter to N.S. Batten, manager, Unemployment Insurance Commission, 4 May 1960, RG 27, Library and Archives Canada.
45 Ibid.
46 Martin, *Once Upon a Mine*, 66–72.
47 Alcan memo, n.d., Alcan Archives Collection.
48 James McGrath to A. Turpin, 4 March 1960, RG 27, Library and Archives Canada,
49 Donald. A. Poynter to N.S. Batten, manager, Unemployment Insurance Commission, 4 May 1960, RG 27, Library and Archives Canada.
50 Ibid.
51 Dembe, *Occupation and Disease*, 1–23.
52 H.A. Etheridge, memo, 10 December 1967, Alcan Archives Collection.
53 Donald. A. Poynter to N.S. Batten, manager, Unemployment Insurance Commission, 4 May 1960, RG 27, Library and Archives Canada.
54 Martin, *Once Upon a Mine*, 66–72.
55 A. Turpin to Walter E. Seibert, 24 January 1958, Joseph R Smallwood Papers, 3:20:088.
56 A.A. Holland to Walter E. Seibert, 2 July 1959, Joseph R. Smallwood Papers, 3:20:087.

CHAPTER SIX

1 Alec T. Hickman, personal communication.

2 Fintan J. Aylward was the son of Patrick Aylward, president of the first trade union in St Lawrence.
3 Aylward, *Report of Royal Commission Respecting Radiation*, 7–14.
4 Slaney, *More Incredible than Fiction.*
5 H.A. Etheridge to Frank deN. Brent, MD, 17 April 1967, Alcan Archives Collection.
6 Aylward, *Report of Royal Commission Respecting Radiation*, 1:273–80, 2:333–5.
7 Ibid., 1:53–65.
8 Ibid.
9 Ibid., 1:67–81.
10 Ibid., 1:83–97.
11 Ibid., 1:221–72.
12 *Newfoundland Gazette*, 20 October 1964.
13 Aylward, *Report of Royal Commission Respecting Radiation*, 1:221–72.
14 Ibid., 1:167–75.
15 Ibid., 1:99–123.
16 Ibid.
17 Ibid., 1:221–72.
18 Ibid.
19 Ibid.
20 Horwood, *Mining as a Way of Death*, 4–7.
21 Leyton, *Dying Hard*, 112–23.
22 Aylward, *Report of Royal Commission Respecting Radiation*, 1:221–72.
23 Ibid.
24 Aylward, *Report of Royal Commission Respecting Radiation*, 1:221–8.
25 Batten, *Report of the Review Committee Workmen's Compensation.*
26 Aylward, *Report of Royal Commission Respecting Radiation*, 1:221–72.
27 Ibid.
28 Ibid.
29 Ibid.; Alec T. Hickman in House of Assembly, Newfoundland and Labrador, 6 April 1970, *Hansard*.
30 Employment Insurance is a striking example of the tendency for government-operated insurance schemes to be both the insurer and social assistance provider. Courchene and Allan, "A Short History of EI and a Look at the Road Ahead."
31 Aylward, *Report of Royal Commission Respecting Radiation*, 1:221–72.
32 Ibid.
33 Ibid.

34 Ibid., 1:99–123.
35 Ibid.
36 Ibid.
37 Ibid.
38 Parsons et al., "Lung Cancer in a Fluorspar Mining Community II," 110–16.
39 Ibid.; deVilliers and Windish, "Lung Cancer in a Fluorspar Mining Community," 94–109.
40 Aylward, *Report of Royal Commission Respecting Radiation*, 1:99–123.
41 Frank G. Barker to Rupert Wiseman, 22 May 1964, Alcan Archives Collection.
42 G. Kane to M.E. Gooding, 7 July 1971, Alcan Archives Collection; G. Kane to Brian Hollywood, 22 July 1974, Alcan Archives Collection.
43 Archer, Saccomanno, and Jones, "Frequency of Different Histologic Types."
44 Saccomanno et al., "Histologic Types of Lung Cancer."
45 Archer, Saccomanno, and Jones, "Frequency of Different Histologic Types."
46 Aylward, *Report of Royal Commission Respecting Radiation*, 1:99–123.
47 Saccomanno et al., "Histologic Types of Lung Cancer."
48 Aylward, *Report of Royal Commission Respecting Radiation*, 1:185–97.
49 Ibid. This was before the introduction of the International Labour Office (ILO) classification of radiographs of pneumoconiosis based on standard comparison films issued by the ILO. These standard films allow for greater consistency in chest x-ray interpretation for the pneumoconioses.
50 Ibid., 1:199–208.
51 Ibid.
52 Ibid.
53 Ibid.
54 Aylward, *Report of Royal Commission Respecting Radiation*, 1:273–80, 2:333–5.
55 Parsons et al., "Lung Cancer in a Fluorspar Mining Community II," 110–16; Aylward, *Report of Royal Commission Respecting Radiation*, 1:209–19.
56 Aylward, *Report of Royal Commission Respecting Radiation*, 1:209–19.
57 Ibid., 1:273–80, 2:333–5, 1:127–31.
58 Aylward, *Report of Royal Commission Respecting Radiation*, 1:273–80, 2:333–5.
59 Aylward, *Report of Royal Commission Respecting Radiation*, 1:133–46.
60 Ibid.
61 Ibid.1:133–46, 1:177–84.
62 Ibid., 1:27–80, 2:333–5.

63 *Decisions of the Government Made in Respect of the Conclusions and Recommendations Set Forth in Ch. XIX of the Royal Commission Respecting Radiation*, Item 33.
64 Aylward, *Report of Royal Commission Respecting Radiation*, 1:43–52.
65 Ibid.
66 Ibid., 1:167–75.
67 Martin, *Leonard Albert Miller*, 119–37.
68 Aylward, *Report of Royal Commission Respecting Radiation*, 1:167–75.
69 *RC Respecting Radiation*, Newfoundland Fluorspar Ltd Submission to the RC Respecting Radiation, brief, Appendix '0,' Box III, 131, i–ii.
70 Gersuny, *Work Hazards and Industrial Conflict*, 20–54.
71 Martin, *Leonard Albert Miller*, 119–37.
72 An Act Respecting the Safety of Workmen in Mines (1951), Amendment #3, 1960, 120 D. *Newfoundland Statutes and Subordinate Legislation*.
73 G. Kane to G.W. Bursey, n.d., Alcan Archives Collection.
74 Frank deN. Brent, MD, memo, 29 June 1967, Alcan Archives Collection.
75 Ibid.
76 H.A. Etheridge, memo, 10 December 1967, Alcan Archives Collection.
77 Rupert Wiseman to Alcan HQ, 11 July 1967, Alcan Archives Collection.
78 Memo, 27 July 1973, Alcan Archives Collection.
79 D. Cant. to G. Kane, 11 May 1973, Alcan Archives Collection.
80 *RC Respecting Radiation*, Newfoundland Fluorspar Ltd Submission to RC Respecting Radiation, testimony of Rupert Wiseman, Box II, 125.
81 *RC Respecting Radiation*, Newfoundland Fluorspar Ltd Submission to RC Respecting Radiation, testimony of Rupert Wiseman, Box II, 125; Martland, "Alcan's Decision," *Evening Telegram*, 12 November 1977.
82 Frank G. Barker, memo, 12 December 1963, Alcan Archives Collection.
83 Ibid.
84 G. Kane to G.W. Bursey, n.d., Alcan Archives Collection.
85 D. Cant to Brian Hollywood, 20 February 1970, Alcan Archives Collection; Frank deN. Brent to Rupert Wiseman and M.E. Gooding, 29 May 1969, Alcan Archives Collection.
86 M.E. Gooding to Frank deN. Brent, 28 May 1969, Alcan Archives Collection.
87 Alice P. Suiker, MD to W.D. Parsons, 4 January 1971, Alcan Archives Collection.
88 Michael P. Slaney, letter to the editor, *Burin Peninsula Post*, 13 January 1973, Alcan Archives Collection.
89 M.E. Gooding and Leonard Albert Miller, 30 December 1970, Alcan Archives Collection.

90 Leonard Albert Miller to M.E. Gooding, 27 February 1971, Alcan Archives Collection.
91 Rupert Wiseman, personal communication.
92 Aylward, *Report of Royal Commission Respecting Radiation*, 2:311–22. The most serious threat to safety was rock bursts. Rock bursts are liable to occur when a hole is drilled into a rock under stress, releasing the compressed tension with explosive force, scattering fragments of rock in all directions. Unfortunately, rock bursts are unpredictable, although at St Lawrence they were more apt to occur in narrow veins.
93 Ibid., 1:7–14.
94 Digest of Newfoundland Fluorspar Ltd, *Submission to the RC Investigating Health and Radiation Condition and Matters Involving Workmen's Compensation*.
95 Aylward, *Report of Royal Commission Respecting Radiation*, 2:311–32.
96 Ibid., 2:295–310.
97 Ibid., 2:283–94.
98 Ibid.
99 Ibid.
100 Ibid.
101 Ibid., 2:295–310.
102 Ibid.
103 Digest of Newfoundland Fluorspar Ltd, *Submission to RC Investigating Health*, 1:89–92.
104 Aylward, *Report of Royal Commission Respecting Radiation*, 2:295–310.
105 Whether the company ever adopted personal dust sampling is not recorded. Memo, 20 July 1976, Alcan Archives Collection; G. Kane, memo, n.d., Alcan Archives Collection.
106 Hickman later broke with Smallwood and crossed the floor of the House. He was minister of justice in Frank Moores's government, which replaced Smallwood's in 1972, and later chief justice.
107 Alec T. Hickman, personal communication.
108 Aylward, *Report of Royal Commission Respecting Radiation*, 1:273–80.
109 Ibid.
110 Ibid.
111 Martin, *Leonard Albert Miller*, 201–3.
112 Hammond, Chaiton, and Collishaw, "Destroyed Documents"; Castleman, *Asbestos*.
113 ACHRE, *Final Report: Advisory Committee on Human Radiation Experiments*, Ch. 12.
114 Ibid.

115 "Be Fair to St Lawrence," *Evening Telegram*, 22 May 1970.
116 Aylward, *Report of Royal Commission Respecting Radiation*, 1:273–80.
117 Ibid.
118 Ibid., 1:67–81.
119 *Decisions of the Government Made in Respect of the Conclusions and Recommendations Set Forth in Ch. XIX of the Royal Commission Respecting Radiation*, Item 14.
120 Rupert Wiseman to M.E. Gooding, 17 March 1970, Alcan Archives Collection.
121 Ibid.
122 Alec T. Hickman in the House of Assembly, Newfoundland and Labrador 6 April 1970, *Hansard*; Slaney, *Evening Telegram*, 20 November 1970.
123 *Decisions of the Government Made in Respect of the Conclusions and Recommendations Set Forth in Ch. XIX of the Royal Commission Respecting Radiation*, Item 14.
124 Ibid.
125 Ibid.
126 Editorial, "Monitors at St Lawrence," *Evening Telegram*, 4 June 1970.
127 Alec T. Hickman in the House of Assembly, Newfoundland and Labrador, 6 April 1970, *Hansard*; "New Safety Measures Planned for St Lawrence Mines," *Evening Telegram*, 14 May 1971.
128 *Evening Telegram*, 26 November 1970.
129 Slaney, *Evening Telegram*, 27 February 1967; W. Callahan, personal communication.
130 Lord Taylor was a physician who had been a UK Labour Party politician. He had acquired a reputation as a labour dispute conciliator by his settlement of the Saskatchewan physicians' strike in the 1960s.
131 W. Callahan, personal communication.
132 Dyer, *Report Number Two*.
133 "Charges of Dangerous Mine Conditions Denied by Alcan," *Daily News*, 19 July 1973.
134 Aylward, *Report of Royal Commission Respecting Radiation*, 1:273–80.
135 Ibid., 1:67–81.
136 Ibid.
137 *Decisions of the Government Made in Respect of the Conclusions and Recommendations Set Forth in Ch. XIX of the Royal Commission Respecting Radiation*, Item 24.
138 Ibid., Item 50.
139 Ibid., Item 64.

140 Ibid., Item 52; Stephen A. Neary in House of Assembly, Newfoundland and Labrador, 6 April 1970, *Hansard.*
141 *Decisions of the Government Made in Respect of the Conclusions and Recommendations Set Forth in Ch. XIX of the Royal Commission Respecting Radiation*, Items 53 and 54.
142 St Lawrence Special Fund, n.d. Joseph R. Smallwood Papers, 3:21:048; A.J. Murphy, ministerial statement, House of Assembly, Newfoundland and Labrador, 8 February 1973, *Hansard.*
143 A.J. Murphy, ministerial statement, House of Assembly, Newfoundland and Labrador, 8 February 1973, *Hansard.*
144 St Lawrence Special Fund, n.d., Joseph R. Smallwood Papers, 3:21:048.
145 G. Kane to J.W. Cameron, 11 December 1970, Alcan Archives Collection.
146 Stephen Neary et al., Memorandum of Meeting, 29 September 1971.
147 Aylward, *Report of Royal Commission Respecting Radiation*, 1:221–72.
148 A.J. Murphy, ministerial statement, House of Assembly, Newfoundland and Labrador, 8 February 1973, *Hansard.*
149 St Lawrence Special Fund, n.d., Joseph R. Smallwood Papers, 3:21:048.
150 Workplace Health, Safety and Compensation Commission of Newfoundland and Labrador (WHSCC), personal communication.
151 Stephen Neary et al., Memorandum of Meeting, 29 September 1971.
152 E. Shainblum, T. Sullivan, and J.W. Frank, "Multicausality, Non-traditional Injury," 58–95.
153 Winter, *Report on the Workman's Compensation Act.*
154 Corlett, *Inspection of Newfoundland Mines for Safety of Workmen and Operations*, 1950; Aylward, *Report of Royal Commission Respecting Radiation*, 1:185–97.
155 Fraser, *Report of the St Lawrence Trade Dispute Board.*
156 Aylward, *Report of Royal Commission Respecting Radiation*, 1:273–80.
157 Ibid., 1:67–81.
158 Ibid., 1:273-80.
159 Ibid., 1:67–81.
160 Ibid.

CHAPTER SEVEN

1 Aylward, *Report of Royal Commission Respecting Radiation*, 1:273–80.
2 *Decisions of the Government Made in Respect of the Conclusions and Recommendations Set Forth in Ch. XIX of the Royal Commission Respecting Radiation*, Item 66.

3 Dyer, *Report Number One.*
4 Workplace Health, Safety and Compensation Commission (WHSCC), personal communication.
5 Ibid.
6 Archer, "Health Concerns in Uranium Mining and Milling," 502–5.
7 Aylward, *Report of Royal Commission Respecting Radiation*, 1:43–52.
8 deVilliers and Windish, "Lung Cancer in a Fluorspar Mining Community," 94–109.
9 Leyton, "The Bureaucratization of Anguish," 70–125.
10 Leyton, *Dying Hard*, 44–55.
11 Martland, "Alcan's Decision Buying Fluorspar Like Buying Pair of Shoes?," *Evening Telegram*, 12 November 1977, 17–19.
12 Leyton, "The Bureaucratization of Anguish," 70–125.
13 Ibid.; Leyton, *Dying Hard*, 44–55.
14 Leyton, "The Bureaucratization of Anguish," 70–125.
15 Leyton, Recommendation 2 in "The Bureaucratization of Anguish," 126–34.
16 Ibid., Recommendation 28.
17 Adams, "The Forgotten Miners," 21–46.
18 William French, "Horror by Horror to Outrage," *Globe and Mail*, 3 May 1975, 36.
19 Memo, 5 May 1975, Alcan Archives Collection.
20 Leyton, *Dying Hard*, 100–11.
21 *Decks Awash*, Special Section, Burin Peninsula, 45–8.
22 *RC Respecting Radiation*, Newfoundland Fluorspar Ltd Submission to RC, brief, Box III, 131, 34–53.
23 *Decks Awash*, Special Section, Burin Peninsula, 45–8.
24 Martland, "Alcan's Decision Buying Flurospar," 17–19.
25 Loder, *Changing Familial Patterns in a Newfoundland Mining Town.*
26 Trist, *The Situation at Newfoundland Fluorspar.*
27 Loder, "Changing Familial Patterns in a Newfoundland Mining Town."
28 Ibid.
29 *RC Respecting Radiation*, Cross-Exam Witnesses, S.T. Payne, Box III, 127.
30 Rennie, *The Dirt*, 108–32.
31 *Decks Awash*, Special Section, Burin Peninsula, 45–8; E. Edwards, personal communication.
32 *St Lawrence Workers Protective Union (CNTU) Submission to the Commission of Inquiry.*
33 Ibid. Another factor that pushed up wages, according to a member of the middle management team, was the bonus system. The system was designed to kick in after a certain production quota had been exceeded

by a worker. Careful thought must be given to setting quotas. If they are set too low, quotas can be costly. This was what was happening at Newfluor; for some workers, bonus payments reached 30 per cent of their wages. Understandably, attempts to raise quota levels or to withdraw bonuses were met with howls of protest from those benefitting from the system.

34 Kealey, *History and Structure of the Newfoundland Labour Movement.*
35 Dyer, *Report Number One.*
36 Strube, *St Lawrence Workers Versus Alcan*, 29–33.
37 Dyer, *Report Number One.*
38 *St Lawrence Workers Protective Union (CNTU) Submission to the Commission of Inquiry.*
39 Dyer, *Report Number One.*
40 Wilkinson, *The Impact of Inequality*, 11, 133.
41 In captive mines it is impossible to come up with exact production costs because extracting, transportation, and refining are parts of the production costs. See Appendix C.
42 Strube, *St Lawrence Workers Versus Alcan*, 29–33.
43 "New Cabinet Initiative to Settle Alcan Dispute," *Evening Telegram*, 24 November 1975.
44 Dyer, *Report Number Two.*
45 Aylward, *Report of Royal Commission Respecting Radiation*, 2:283–94.
46 Strube, *St Lawrence Workers Versus Alcan*, 29–33.
47 *Hansard*, 18 May 1976.
48 Martland, "Alcan's Decision Buying Fluorspar," 17–19; P. Cooper, personal communication. *Daily News*, 26 June 1977.
49 Martland, "Alcan's Decision Buying Fluorspar," 17–19.
50 Ibid.
51 Dyer, *Report Number One.*
52 P. Cooper, personal communication.
53 Ibid.
54 Rennie, *The Dirt*, 108–32.
55 Editorial, "The St Lawrence Story," *Daily News*, 22 June 1977.
56 "NFLD Fluorspar Mines Will Close Next Year," *Northern Miner*, 23 July 1977.
57 "Time Has Run Out for Fluorspar Mining Town," *Globe and Mail*, 30 July 1977.
58 Martland, "Alcan's Decision Buying Fluorspar," 17–19.
59 Hodge and Partners, *Report on an Appraisal of the Alcan Production Operations.*
60 Duncan C. Campbell, Global Mission, 1232–4, Alcan Archives Collection.

61 "NFLD Fluorspar Mines Will Close Next Year," *Northern Miner*, 23 July 1977.
62 *Compass*, 6 April 1978, Alcan Archives Collection.
63 "Time Has Run Out for Fluorspar Mining Town," *Globe and Mail*, 30 July 1977.
64 Brian Peckford qtd. in Martland, "Alcan's Decision Buying Fluorspar," 17–19.
65 Ed Broadbent qtd. in "Support to Keep Mines Open in Newfoundland is Pledged," *Halifax Chronicle Herald*, 15 October 1977; Rodriguez, "Fluorspar Tariffs Urged," *Hamilton Spectator*, 2 February 1978.
66 Hon. John C. Crosbie, House of Commons debates, 7 February 1978, *Hansard*.
67 Alec T. Hickman qtd. in Martland, "Alcan's Decision Buying Fluorspar," 17–19.
68 Alec T. Hickman qtd. in "Ottawa Ignored Town's Plight, Minister Says," *Hamilton Spectator*, 26 October 1977.
69 "Quotas Ruled Out," *Winnipeg Tribune*, 14 October 1977.
70 "Alcan Offers Fluorspar Mines to Newfoundland for $1.00 A Year," *Montreal Gazette*, 8 December 1977.
71 Ibid.
72 Ibid.
73 "Claim St Lawrence Assets Worth $10 Million: Alcan Wants Fair Value," *Evening Telegram*, 19 December 1977.
74 Ibid.
75 "$1.00 Lease of Mine Rejected by NFLD," *Edmonton Journal*, 16 December 1977.
76 Ibid.
77 "Proposal to Alcan," *Globe and Mail*, 23 December 1977.
78 "Claim St Lawrence Assets Worth $10 Million, Alcan Wants Fair Value," *Evening Telegram*, 19 December 1977.
79 An Act Relating to Mines and Quarries, *Newfoundland Statutes and Subordinate Legislation*.
80 Martin, *Once Upon a Mine*, 66–72.
81 Ron Dawe, "Reasons for Hope St Lawrence Mine Will Reopen," *Newfoundland Herald*, 19 November 1983, 18–19.
82 Martland, "Alcan's Decision Buying Fluorspar," 17–19.
83 Ibid.
84 Ibid.; P. Cooper, personal communication.
85 Martland, "Alcan's Decision Buying Fluorspar," 17–19.
86 Hon. Jerome W. Dinn, ministerial statement, 26 June 1981.

87 Hon. T.A. Blanchard, ministerial statement, 28 May 1986.
88 Ibid.
89 Martland, "Alcan's Decision Buying Fluorspar," 17–19.
90 Ibid.
91 "At the Top of the Great Bauxite Mountain," *Financial Post*, 7 January 1978.
92 Martland, "Alcan's Decision Buying Fluorspar," 17–19.
93 Ibid.
94 "Pullout in St Lawrence: It's Not a Unique Situation," *Evening Telegram*, 12 November 1977, 17.

CHAPTER EIGHT

1 Newfoundland Information Services (NIS), Executive Council, 11 June 1984, Legislative Library.
2 Hodge and Partners, *Report on Appraisal of the Alcan Production Operations*.
3 "Once Hazardous Newfoundland Mine Offers Hope," *London Free Press*, 30 March 1985.
4 News Release, St Lawrence Fluorspar Company Ltd, Government of Canada, 16 August 1984, Legislative Library.
5 Ernst and Young, *The Assets and Operations of St Lawrence Fluorspar Ltd.*
6 Ibid.
7 Murray, "British Firm Planning to Reopen Former Alcan Producers," 47–50.
8 Ibid.
9 News Release, St Lawrence Fluorspar Company Ltd, Government of Canada, 16 August 1984, Legislative Library.
10 Ibid.
11 Hon A. Brian Peckford, statement, 26 March 1985, Legislative Library.
12 Ibid.
13 News Release, St Lawrence Fluorspar Company Ltd, Government of Canada, 16 August 1984, Legislative Library.
14 "None of Their Business Says Mayor: Labour Raps Government Council for Concessions to British Mining Company," *Evening Telegram*, 1 November 1985.
15 Ibid.
16 Noonan, "Colonial Traditions, British Mining in Newfoundland," 14–15.
17 "None of Their Business Says Mayor," *Evening Telegram*, 1 November 1985.
18 Ibid.

19 Ernst and Young, *The Assets and Operations of St Lawrence Fluorspar Ltd.*; G. Doyle, personal communication.
20 G. Doyle, personal communication.
21 Briefing Note, St Lawrence Flourspar (*sic*) Mine, n.d., Legislative Library, House of Assembly, NL.
22 Ibid.
23 Ibid.
24 Story, "Newfoundland Fluorspar Miners."
25 Linda Strowbridge, "Government Guarantée Prevents Layoffs," *Sunday Express*, 27 May 1987.
26 "Welcome Christmas Present for Two Provincial Mines," *Evening Telegram*, 28 December 1987.
27 Ibid.; "St Lawrence Mine Layoff Scrapped, n.d., n.p., date stamped 23 May 1987, Legislative Library, House of Assembly, NL.
28 Briefing Note, St Lawrence Flourspar (*sic*) Mine, n.d., Legislative Library, House of Assembly, NL.
29 *Evening Telegram*, 28 December 1987.
30 Briefing Note, St Lawrence Flourspar (*sic*) Mine, n.d., Legislative Library, House of Assembly, NL.
31 In the late 1970s, all government occupational health and safety monitoring and health services were consolidated into the Division of Occupational Health and Safety, a division of the Department of Labour.
32 Noonan, "Colonial Traditions, British Mining in Newfoundland," 14–15; Story, "Newfoundland Fluorspar Miners," 19–21.
33 "Province Forbids Former Miners From Working at Fluorspar Mine," *Sunday Express*, 25 September 1988.
34 Roger March, personal communication.
35 G. Doyle, personal communication; Michael Harris and Linda Strowbridge, "St Lawrence Mislead on Mine Pollution Documents," *Sunday Express*, 24 April 1988.
36 "Official Denies Own Report Misled St Lawrence on Mine Pollution, Documents Show," *Sunday Express*, 24 April 1988.
37 Ibid.
38 Ibid.
39 "Province Forbids Foremen Miners From Working at Fluorspar Mine," *Sunday Express*, 25 September 1988.
40 Ernst and Young, *The Assets and Operations of St Lawrence Fluorspar Ltd.*; Hon. Rex Gibbons, minister of mines and energy, ministerial statement, 10 May 1994, Legislative Library.

41 Hon. Rex Gibbons, minister of mines and energy, ministerial statement, 10 May 1994, Legislative Library.
42 Ibid.
43 Ibid.
44 Ernst and Young, *The Assets and Operations of St Lawrence Fluorspar Ltd.*
45 P. Cooper, personal communication.
46 Hodge, *Report on Appraisal of the Alcan Production Operations.*
47 Diamond, *Collapse,* 441–85.
48 Ibid.
49 Sweet, "St Lawrence Fluorspar Operation Gets Loan," *Telegram* (St John's), 12 August 2011.
50 Morrison et al., "Cancer Mortality Experience," 1266–75; Morrison et al., "Radon-Progeny Exposure," 58–65; Villeneuve, Morrison, and Lane, "Radon and Lung Cancer Risk," 157–69.
51 Villeneuve, Morrison, and Lane, "Radon and Lung Cancer Risk," 157–69.
52 Ibid.
53 Ibid.
54 Morrison et al., "Radon-Progeny Exposure," 58–65.
55 Wilkinson, *The Impact of Inequality*, 114, 128–30, 139.
56 Ibid.
57 Villeneuve, Morrison, and Lane, "Radon and Lung Cancer Risk," 157–69.
58 Morrison et al., "Cancer Mortality Experience," 1266–75; Morrison et al., "Radon-Progeny Exposure," 58–65; Villeneuve, Morrison, and Lane, "Radon and Lung Cancer Risk," 157–69.
59 Morrison et al., "Radon-Progeny Exposure," 58–65; Villeneuve, Morrison, and Lane, "Radon and Lung Cancer Risk," 157–69.
60 Ibid.
61 Villeneuve, Morrison, and Lane, "Radon and Lung Cancer Risk," 157–69.
62 Ibid.
63 Ibid.
64 Letourneau qtd. in Villeneuve, Morrison, and Lane, "Radon and Lung Cancer Risk," 157–69.
65 Ibid.
66 Villeneuve, Morrison, and Lane, "Radon and Lung Cancer Risk," 157–69.
67 Aylward, *Report of Royal Commission Respecting Radiation*, 1:99–123.
68 Morrison et al., "Radon-Progeny Exposure," 58–65.
69 Villeneuve, Morrison, and Lane, "Radon and Lung Cancer Risk," 157–69.
70 Ibid.
71 Ibid.

72 Ibid.
73 Ibid.
74 Ibid.; Saccomanno and Archer qtd. in Villeneuve, Morrison, and Lane, "Radon and Lung Cancer Risk," 157–69; Wright and Couves, "Radiation Induced Carcinoma of the Lung," 495–8.
75 Villeneuve, Morrison, and Lane, "Radon and Lung Cancer Risk," 157–69.
76 Morrison et al., "Radon-Progeny Exposure," 58–65.
77 Villeneuve, Morrison, and Lane, "Radon and Lung Cancer Risk," 157–69.
78 Morrison et al., "Radon-Progeny Exposure," 58–65.
79 Morrison et al., "Cancer Mortality Experience," 1266–75. The present ACGIH recommended radon and radon progeny permissible exposure level is still 4.0 WLMS, the one currently followed in Newfoundland and Labrador.

Bibliography

ARCHIVAL MATERIALS

Alcan Archives Collection. Rio Tinto Alcan. Montreal, QC.

Aylward, Fintan J. *Report of Royal Commission Respecting Radiation Compensation and Safety at the Fluorspar Mines St Lawrence Newfoundland.* Sections 1 and 2. July 1969. Legislative Library, House of Assembly, NL.

Batten, H.M. *Report of the Review Committee Workmen's Compensation Newfoundland and Labrador.* July 1972. Archives and Special Collections, Memorial University of Newfoundland.

Blanchard, Hon. T.A. Ministerial Statement, 28 May 1986. Legislative Library, House of Assembly, NL.

Canada. *House of Commons Debates (Hansard).*

Canada Tariff Board. *Report by the Tariff Board Relative to the Investigation Ordered by the Minister of Finance Respecting Fluorspar.* 10 September 1958. Centre for Newfoundland Studies, Memorial University of Newfoundland.

Commission of Government, Secretary of the Commission Fonds. PANL GN-38. The Rooms Corporation of Newfoundland and Labrador, Provincial Archives Division.

Corlett, A.V. *The Safety of Workmen in Newfoundland Mining Operations.* 1948. Commission of Government, Secretary of the Commission Fonds. GN38 S5-4-1, file 1. The Rooms Corporation of Newfoundland and Labrador, Provincial Archives Division.

– *Inspection of Newfoundland Mines for Safety of Workmen and Operations.* 1949. Centre for Newfoundland Studies, Memorial University of Newfoundland.

– *Inspection of Newfoundland Mines for Safety of Workmen and Operations*. 1950. The Rooms Corporation of Newfoundland and Labrador, Provincial Archives Division.

Decisions of the Government Made in Respect of the Conclusions and Recommendations Set Forth in Ch. XIX of the Royal Commission Respecting Radiation, Compensation, and Safety at the St Lawrence Fluorspar Mines. n.d. Legislative Library, House of Assembly, NL.

Department of Labour Strikes and Lockouts File. RG 27, vol. 540, reel T3401. Department of Labour Fonds. Library and Archives Canada, Ottawa, ON.

Department of Natural Resources Fonds. PANL, GN 31/3A. The Rooms Corporation of Newfoundland and Labrador, Provincial Archives Division.

Dinn, Hon. Jerome W. Ministerial Statement, 26 June 1981. Legislative Library, House of Assembly, NL.

Dyer, Howard J. *Report Number One of the Industrial Inquiry Commission into the Dispute Between the Aluminium Company of Canada Limited St Lawrence and the St Lawrence Workers Protective Union (CNTU)*. St Lawrence, NL. 18 August 1975. Legislative Library, House of Assembly, NL.

– *Report Number Two of the Industrial Inquiry Commission into Safety Conditions at the Mining Operations of the Aluminium Company of Canada Limited*. St Lawrence, NL. 23 November 1976. Legislative Library, House of Assembly, NL.

Ernst and Young. *The Assets and Operations of St Lawrence Fluorspar Ltd*. 1991. Centre for Newfoundland Studies, Memorial University of Newfoundland.

Fraser, A.M. *Report of the St Lawrence Trade Dispute Board (TDB). Appointed under order of Commissioner for Public Utilities. St John's: 1941. St Lawrence Mining Dispute, 1940–1942*. Commission of Government, Secretary of the Commission Fonds. GN-38 S5-4-1 File 5. The Rooms Corporation of Newfoundland and Labrador, Provincial Archives Division.

Gibbons, Hon. Rex. Ministerial Statement (Mines and Energy), 10 May 1994. Legislative Library, House of Assembly, NL.

Higgins, Gordon F. *Report of the Mining Industry of Newfoundland to the National Convention, Attachment: General Information on the History of the Activities of the Geological Survey of Newfoundland by C.K. Howse*. 26 September 1946. Commission of Government, Secretary of the Commission Fonds. GN-38 S5-4-1 File 2. The

Rooms Corporation of Newfoundland and Labrador, Provincial Archives Division.

Hodge, B.L. and Partners. *Report on an Appraisal of the Alcan Production Operations Based on St Lawrence Fluorspar Deposits Newfoundland for the Government of Newfoundland.* November 1977. Legislative Library, House of Assembly, NL.

Howse, C.K. "Our Story." Unpublished autobiography. 1993. Archives and Special Collections, Memorial University of Newfoundland.

Loder, Richard L. "Changing Familial Patterns in a Newfoundland Mining Town: A Town of Widows and Orphans." Paper presented to Professor Louis Charimonte. 10 November 1972. Centre for Newfoundland Studies, Memorial University of Newfoundland.

Neary, Stephen A. et al. Memorandum of Meeting, 29 September 1971. Legislative Library, House of Assembly, NL.

Newfoundland Fluorspar Ltd. *Submission to the RC Investigating Health and Radiation Condition and Matters Involving Workmen's Compensation, Rock Stresses, Rock Bursts, Mining Methods, Company Safety Policy and Practice, Accident and Safety Training.* Parts 1 and 2. April 1968. St Lawrence, NL. Legislative Library, House of Assembly, NL.

Newfoundland and Labrador. *House of Assembly Debates (Hansard).*

– *Newfoundland Statutes and Subordinate Legislation, Consolidated to May 21, 1998.* Legislative Library, House of Assembly, NL.

– Royal Commission to Investigate Health and Radiation Conditions as well as matters involving the Payment of Workmen's Compensation to Disabled Workmen and Dependents of Deceased Workmen Employed in the Operations of the Fluorspar Mines at St Lawrence. *Royal Commission (RC) Respecting Radiation, Compensation and Safety at the Fluorspar Mines, St Lawrence, Newfoundland, 1969.* GN 139 R6-E-2, Boxes II and III. The Rooms Corporation of Newfoundland and Labrador, Provincial Archives Division.

– *The Revised Statutes of Newfoundland, 1952: A Revision and Consolidation of the Public General Statutes of Newfoundland as Contained in the Consolidated Statutes (Third Series) and as Passed in the years 1917 to 1952 both Inclusive.* Legislative Library, House of Assembly, NL.

Newfoundland Information Services (NIS). Executive Council, 11 June 1984. Legislative Library, House of Assembly, NL.

Patterson, T.H. NHW Occupational Health Division to Leonard Albert Miller, 30 December 1959, Department of National Health and Welfare, Canada. Dr John Martin Fonds, Coll-014. Faculty of Medicine Founders' Archive, Memorial University of Newfoundland.

Peat, Marwick, Mitchell and Co. *St Lawrence Corporation of Newfoundland Limited and St Lawrence Fluorspar Inc: A Report on Application for Financial Assistance*. 1950. Centre for Newfoundland Studies, Memorial University of Newfoundland.

Peckford, Hon A. Brian. Statement, 26 March 1985. Legislative Library, House of Assembly, NL.

Seibert, Walter E. *Brief Supporting a Tariff Request on Fluorspar Entering Into Canada. Submitted to the Tariff Board of Canada*. 1958. Centre for Newfoundland Studies, Memorial University of Newfoundland.

Smallwood, Joseph R., Papers. Coll-075. Archives and Special Collections, Memorial University of Newfoundland.

Snelgrove, A.K. *Mines and Mineral Resources of Newfoundland: Geological Survey*. Map. St John's, NL. 1938. Centre for Newfoundland Studies, Memorial University of Newfoundland.

St Lawrence Corporation and Newfoundland Fluorspar Limited. RG 39–1, vol. 38, file 35–02-0. Library and Archives Canada, Ottawa, ON.

St Lawrence Workers Protective Union (CNTU) Submission to the Commission of Inquiry into the Industrial Dispute at the Alcan Fluorspar Works in St Lawrence Newfoundland. Presented by the St Lawrence Workers Protective Union. Confederation of National Trade Unions Economic Research Department. 13 August 1975. Legislative Library, House of Assembly, NL.

Strube, Martin, and Alternate Press Staff. "St Lawrence Workers Versus Alcan" *Alternate Press* (Toronto), August 1971. Legislative Library, House of Assembly, NL.

Windish, J.P., and H.P. Sanderson. "Dust Hazards in the Mines of Newfoundland: I. Newfoundland Fluorspar Limited, St Lawrence." Laboratory Services, Occupational Health Division, Department of National Health and Welfare, Ottawa. 1958. Centre for Newfoundland Studies, Memorial University of Newfoundland.

Winter, H.A. *Report on the Workman's Compensation Act. Report of the Review Committee Appointed to Review, Consider, Report Upon, and Make Recommendations Respecting the Workmen's Compensation Act*. 9 May 1966. Centre for Newfoundland Studies, Memorial University of Newfoundland.

SECONDARY SOURCES

Adams, Ian. "The Forgotten Miners," *MacLeans* (June 1967) 21–46.

Advisory Committee on Human Radiation Experiments (ACHRE). *Final Report: Advisory Committee on Human Radiation Experiments*. US Government Printing Office: Washington, DC, 1995.

Archer, V.E. "Health Concerns in Uranium Mining and Milling." *Journal of Occupational Medicine* 23 (1981): 502–5.

Archer, V.E., G. Saccomanno, J.H. Jones, "Frequency of Different Histologic Types of Bronchiogenic Carcinoma as Related to Radiation Exposure." *Cancer* 34 (1974): 2056–60.

Bassler, G.P., *Alfred Valdmanis and the Politics of Survival*. Toronto: University of Toronto Press, 2000.

Castleman, Barry I. *Asbestos: Medical and Legal Aspects*. New York: Harcourt Brace Jovanovich Publishers, 1984.

Courchene, T.J., and J.R. Allan, "A Short History of EI and a Look at the Road Ahead." *Policy Options* 30, no. 8 (2009): 19–28.

Decks Awash. Special Section, Burin Peninsula. 4, no. 6 (December 1975): 45–8.

Dembe, A.E. *Occupation and Disease: How Social Factors Affect the Conception of Work Related Disorders*. New Haven: Yale University Press, 1996.

deVilliers, A.J., and J.P. Windish. "Lung Cancer in a Fluorspar Mining Community, I Radiation, Dust, and Mortality Experience." *British Journal of Industrial Medicine* 21 (1964): 94–109.

Diamond, Jared. *Collapse: How Societies Choose to Fail or Succeed*. London: Penguin Books, 2005.

Doll, R., and A.B. Hill. "The Mortality of Doctors in Relation to Their Smoking Habits: A Preliminary Report." *British Medical Journal* 1 (1954): 1451–5.

Editorial. "New Fluorspar Developments on the Burin Peninsula, Newfoundland." *Journal of Commerce* 20, no. 5 (1953): 23.

Edwards, Ena Farrell, and R.E. Buehler. *Notes Toward a History of St Lawrence*. St John's, NL: Breakwater Books, 1967.

Edwards, Gordon. *Health Effects of Ionizing Radiation: A Comprehensive Report from the Canadian Environmental Advisory Council on the Biological Effects Of Radiation, Genetic Damage, Latency of Cancer*. Canadian Coalition for Nuclear Responsibility. N.d. Accessed 1 December 2008. www.ccnr.org/ceac_B.html_62K.

Fishback, Price V., and Shawn Everett Kantor. *A Prelude to the Welfare State: The Origins of Workers' Compensation*. Chicago: University of Chicago Press, 2000.

Fogwill, Irving. "How Modern Workmen's Compensation Law Came to Newfoundland." N.d. Courtesy of WHSCC.

Galbraith, John Kenneth. *The New Industrial State*. New York: Signet, 1968.

Gersuny, Carl. *Work Hazards and Industrial Conflict*. Hanover, NH: University Press of New England, 1981.

Gillespie, Bill. *A Class Act: An Illustrated History of the Labour Movement in Newfoundland and Labrador*. St John's, NL: Newfoundland and Labrador Federation of Labour, 1986.

Gwyn, Richard. *Smallwood: The Unlikely Revolutionary*. Toronto: McClelland and Stewart, 1968.

Hammond, D., M. Chaiton, A. Lee, and N. Collishaw, "Destroyed Documents: Uncovering the Science that Imperial Tobacco Canada Sought to Conceal." *Canadian Medical Association Journal* 181 (2009): 691–8.

Hattenhauer, R. *A Brief Labour History of Newfoundland*. St John's, NL, 1970.

Hillgartner, Stephen. "The Political Language of Risk: Defining Occupational Health." In *The Language of Risk: Conflicting Perspectives on Occupational Health*, edited by Dorothy Nelkin. Beverley Hills, CA: Sage Publications, 1985.

Horwood, Harold. "Mining as a Way of Death: Human Life and Radiation in St Lawrence, Newfoundland." *Weekend Magazine* (St John's, NL), 31 May 1975.

Howse, A., P. Dean, S. Swindon, B. Kean, and F. Morrisey. *Fluorspar Deposits in the St Lawrence Area, Newfoundland Geology and Economic Potential*. St John's, NL: Department of Mines and Energy, Mineral Development Division, 1983.

Kealey, Gregory S. *The History and Structure of the Newfoundland Labour Movement*. Royal Commission on Employment and Unemployment Newfoundland and Labrador, 1986.

Leyton, Elliott. "The Bureaucratization of Anguish: The Workmen's Compensation Board in an Industrial Disaster." In *Bureaucracy and World View: Studies in the Logic of Official Interpretation*, by Don Handelman and Elliott Leyton, 70–125. St John's, NL: Institute of Social and Economic Research, Memorial University of Newfoundland, 1978.

– *Dying Hard: The Ravages of Industrial Carnage*. 1975. Toronto: McClelland and Stewart, 1990.

– "Recommendations (Appendix)," in *Bureaucracy and World View: Studies in the Logic of Official Interpretation*, by Don Handelman and Elliott Leyton, 126–34. St John's, NL: Institute of Social and Economic Research, Memorial University of Newfoundland, 1978.

Liddell, Thomas K. *Industrial Survey of Newfoundland, September–October 1938, April–September 1939*. St John's, NL: Robinson and Co. Ltd, 1940.

Martin, John R. *Leonard Albert Miller, Public Servant*. Markham, ON: Associated Medical Services Inc/The Hannah Institute for the History of Medicine; Fitzhenry and Whiteside Ltd, 1998.

Martin, Wendy. *Once Upon a Mine: Story of Pre-Confederation Mines on the Island of Newfoundland*. Special Vol. 26. The Canadian Institute of Mining and Metallurgy. Ste Anne-de-Bellevue, QC: Harper's Press Cooperative, 1983.

Martland, Sandra. "Alcan's Decision: Buying Fluorspar Like Buying a Pair of Shoes." *Evening Telegram*, 12 November 1977, 17–19.

Morrison, H.I., R.M. Semenciw, Y. Mao, D.T. Wigle. "Cancer Mortality Experience Among a Group of Fluorspar Miners Exposed to Radon Progeny." *American Journal of Epidemiology* (1988) 128: 1266–75.

Morrison, H.I., P.J. Villeneuve, J.H. Lubin, and D.E. Schaube. "Radon-Progeny Exposure and Lung Cancer Risk in a Cohort of Newfoundland Fluorspar Miners." *Radiation Research* 150, no. 1 (1998): 58–65.

Muir, D.C.F., H.S. Shannon, J.A. Julian, D.K. Verma, A. Sebestyen, and C.D. Bernholz. "Silica Exposure and Silicosis among Ontario Hard Rock Miners. I Methodology." *American Journal of Industrial Medicine* 16 (1989): 5–11.

Murray, Jane. "British Firm Planning to Reopen Former Alcan Producers." *Canadian Mining Journal* 105, no. 10 (1984): 47–50.

Neary, Peter F. *Newfoundland in the North Atlantic World, 1929–1949*. Montreal and Kingston: McGill-Queen's University Press, 1988.

Noonan, Susan. "Colonial Traditions, British Mining in Newfoundland." *Our Times* 5, no. 6 (1986): 14–15.

Parsons, W.D., A.J. deVilliers, L.S. Bartlett, and M.R. Becklake. "Lung Cancer in a Fluorspar Mining Community II, Prevalence of Respiratory Symptoms and Disability" *British Journal of Industrial Medicine* 21 (1964): 110–16.

"Pullout in St Lawrence: It's Not a Unique Situation." *Evening Telegram*, 12 November 1977, 17.

Rennie, Rick. *The Dirt: Industrial Disease and Conflict at St Lawrence, Newfoundland*. Halifax and Winnipeg: Fernwood Publishing, 2008.

– "The Historical Origins of an Industrial Disaster: Occupational Health and Labour Relations at the Fluorspar Mines, St Lawrence, Newfoundland, 1933–1945." *Labour/Le Travail* 55 (2005): 107–42.

– "The St Lawrence Fluorspar Mines: A Brief History." *Newfoundland and Labrador Heritage*. 1998. http://www.heritage.nf.ca/society/stlawrence.html.

Rosen, George. *The History of Miners' Diseases: A Medical and Social Interpretation*. New York: Schuman's, 1943.

Rosner, David, and Gerald Markowitz. *Deadly Dust: Silicosis and the On-Going Struggle to Protect Workers' Health*. Princeton: Princeton University Press, 1991.

Saccomanno, E., V.E. Archer, O. Auerbach, M. Kuschner, R.P. Saunders, and M.G. Klein. "Histologic Types of Lung Cancer Among Uranium Miners." *Cancer* 27 (1971): 515–23.

Sagoff, Mark. "Sense and Sentiment in Occupational Safety and Health." In *The Language of Risk: Conflicting Perspectives on Occupational Health*, edited by Dorothy Nelkin, 179–97. Beverley Hills, CA: Sage Publications, 1985.

Shainblum, E., T. Sullivan, and J.W. Frank. "Multicausality, Non-traditional Injury, and the Future of Workers' Compensation." In *Workers' Compensation: Foundation for Reform*, edited by M. Gunderson and D. Hyatt, 58–95. Toronto: University of Toronto Press, 2000.

Sigerist, Henry. Introduction to *The History of Miners' Diseases: A Medical and Social Interpretation* by George Rosen, ix–xii. New York: Schuman's, 1943.

Slaney, Rennie. *More Incredible than Fiction. The True Story of the Indomitable Men and Women of St Lawrence, Newfoundland from the Time of Settlement to 1965: History of Fluorspar Mining at St Lawrence Newfoundland/As told by one of the Miners*. Montreal: La Confederation des syndicats nationaux, 1975.

"St Lawrence to Point May." *Decks Awash* 13, no. 6 (November–December 1984): 5–12.

Story, Alan. "Newfoundland Fluorspar Miners, Dying With Money or Living Without." *Canadian Dimension* 20, no. 2 (1986): 19–21.

Sweet, Barb. "St Lawrence Fluorspar Operation Gets Loan." *Telegram* (St John's), 12 August 2011.

Thompson, E.P. *The Making of the English Working Class*. 1963. New York: Penguin Books, 1991.

– "Time, Work-Disciplining and Industrial Capitalism." In *Essays in Social History*, edited by M.W. Flinn and T.C. Smout, 56–97. London: Oxford University Press, 1974.

Trist, Eric L. "The Situation at Newfoundland Fluorspar Ltd in Relation to the St Lawrence Community: A Social Science Appreciation Based on a Preliminary Field Trip Visit During the Week of March 18-23, 1968." Courtesy of Rick Rennie.

Villeneuve, P.J., H.J. Morrison, and R. Lane. "Radon and Lung Cancer Risk: An Extension of the Mortality Follow-up of the Newfoundland Fluorspar Cohort." *Health Physics* 92, no. 2 (2007): 157–69.

Wardle, R.J. "The Mineral Industry in Newfoundland and Labrador: Its Development and Economic Contributions." Open File NFLD/2889. Government of Newfoundland and Labrador, Department of Natural Resources Geological Survey, St John's, NL, October 2004.

Weir, Gail. *The Miners of Wabana*. 2nd ed., St John's, NL: Breakwater Books Ltd, 2006.

Wilkinson, Richard G. *The Impact of Inequality: How to Make Sick Societies Healthier*. New York: New Press, 2005.

Wright, E.S., and C.M. Couves. "Radiation Induced Carcinoma of the Lung. The St Lawrence Tragedy." *Journal of Thoracic and Cardiovascular Surgery* 74 (1977): 495–8.

INTERVIEWS AND PERSONAL COMMUNICATION

Arnold, Dr Ian	
Bartlett, Eric	WHSCC
Callahan, William	Minister of Mines
Cooper, Phonce	Engineer Newfluor and St Lawrence Fluorspar Co. Ltd
deVilliers, Dr Arnold*	Epidemiologist NHW
Doyle, George	St Lawrence Fluorspar Co. Ltd
Drover, Gerald	Safety Officer, Newfluor
Edwards, Ena*	Historian, St Lawrence
Etchegary A.A.	Former Resident of St Lawrence
Goodyear, David	Dr H. Bliss Centre, Radiologist and Oncologist
Hearl, Frank J.	NIOSH
Hickman, T. Alec	Hon. Chief Justice
Hollywood, Dr Brian*	Physician, St Lawrence
Kane, Dr Gerald*	Physician, Alcan
March, Roger	Chief Inspector of Mines

Martin, Dr Christopher J.	Occupational Health Physician
Muir, Dr David	McMaster University, Occupational Health
Parsons, Dr W.D.	Physician
Pike, Levi	Miner
Poynter, Adele	Donald A. Poynter's daughter
Slaney, Herbert	Laboratory Technician, Newfluor
Turpin, Gerald	Paymaster, Newfluor
Walsh, Cyril J.*	Cottage Hospital Doctor
Wiseman, Rupert	Manager, Newfluor
Young, Dr Peter	Mining Engineer

* Deceased

Index